Specialty Photonic Crystal Fibres and Their Applications

Specialty Photonic Crystal Fibres and Their Applications

Editors

David Novoa

Nicolas Y. Joly

MDPI • Basel • Beijing • Wuhan • Barcelona • Belgrade • Manchester • Tokyo • Cluj • Tianjin

Editors
David Novoa
Department of Communications Engineering, School of Engineering, University of the Basque Country (UPV/EHU)
IKERBASQUE, Basque Foundation for Science
Bilbao
Spain

Nicolas Y. Joly
Max-Planck Institute for the Science of Light and Department of Physics
Friedrich-Alexander University Erlangen
Germany

Editorial Office
MDPI
St. Alban-Anlage 66
4052 Basel, Switzerland

This is a reprint of articles from the Special Issue published online in the open access journal *Crystals* (ISSN 2073-4352) (available at: www.mdpi.com/journal/crystals/special_issues/Photonic_Crystal_Fibres).

For citation purposes, cite each article independently as indicated on the article page online and as indicated below:

LastName, A.A.; LastName, B.B.; LastName, C.C. Article Title. *Journal Name* **Year**, *Volume Number*, Page Range.

ISBN 978-3-0365-1614-1 (Hbk)
ISBN 978-3-0365-1613-4 (PDF)

Contents

Editorial

Specialty Photonic Crystal Fibers and Their Applications

David Novoa [1,2,3,*] and Nicolas Y. Joly [3,4]

1 Department of Communications Engineering, Engineering School of Bilbao, University of the Basque Country (UPV/EHU), Torres Quevedo 1, 48013 Bilbao, Spain
2 IKERBASQUE, Basque Foundation for Science, Plaza Euskadi 5, 48009 Bilbao, Spain
3 Max-Planck Institute for the Science of Light, Staudtstrasse 2, 91058 Erlangen, Germany; nicolas.joly@fau.de
4 Department of Physics, Friedrich-Alexander-Universität, Staudtstrasse 2, 91058 Erlangen, Germany
* Correspondence: david.novoa@ehu.eus

Citation: Novoa, D.; Joly, N.Y. Specialty Photonic Crystal Fibers and Their Applications. *Crystals* **2021**, *11*, 739. https://doi.org/10.3390/cryst11070739

Received: 22 June 2021
Accepted: 23 June 2021
Published: 25 June 2021

This year not only commemorates the 60th anniversary of nonlinear optics with the seminal experiment of second harmonic generation [1], but it is also the 30th anniversary of the invention of the photonic crystal fiber (PCF) [2]. Following their first practical demonstration in 1996 [3], PCFs [4,5] have rapidly evolved into an established platform for applications in both academic and industrial environments. Their unique ability to confine light in a far more versatile way than possible with conventional optical fibers facilitated the expansion of the multifaceted world of PCF to cover not only nonlinear optics [6], but also many other disparate fields such as interferometry [7], beam delivery [8], laser science [9], telecommunications [10], quantum optics [11], sensing [12,13], microscopy [14], and many others.

More recently, there has been a great interest in the design, fabrication, and application of specialty PCFs with otherwise inaccessible and exotic properties far beyond the capabilities of standard fibers. These include, but are not limited to, the close-to-real fulfilment of the long-standing dream of lowering the minimum attenuation achievable in current fibers for communications [15], the use of glasses other than silica to manufacture PCFs with an enhanced performance in originally forbidden spectral regions such as the ultraviolet (UV) [16] or the mid-infrared (MIR) [17], or the three-dimensional engineering of fibers to open new horizons in fundamental science [18,19]. Furthermore, specialty PCFs have also jumped into real-world applications, playing a key role in the development of, e.g., the new generation of high-power fiber lasers [20] or constituting a new paradigm in photochemistry [21,22].

This special issue in Crystals is intended to provide an overview of the state-of-the-art in specialty PCF technology and its multiple applications, combined with an optimistic outlook to what lies ahead. It comprises six original research papers and one review from different leading research institutions worldwide.

The review Application of Hollow-Core Photonic Crystal Fibers in Gas Raman Lasers Operating at 1.7 μm [23] by the Changsha Environmental Protection Vocational College, the National University of Defense Technology, and the State Key Laboratory of Pulsed Power Laser Technology (China) focuses on the detailed description of the origin and development of near-infrared narrowband pulsed laser sources based on stimulated Raman scattering in gas-filled hollow-core PCFs. It includes a thorough revision of the literature on this technology, which has successfully been applied through the years in many spectral regions from the UV [24,25] to the visible [26,27] and infrared [28,29]. In addition, the research paper Hydrogen Molecules Rotational Stimulated Raman Scattering in All-Fiber Cavity Based on Hollow-Core Photonic Crystal Fibers [30] by the same group reports on the implementation of a laser cavity to enhance rotational SRS in a hydrogen-filled hollow-core PCF.

The original research paper *Geometrical Scaling of Antiresonant Hollow-Core Fibers for Mid-Infrared Beam Delivery* [31] by the *Nanyang Technological University* (Singapore) reports a detailed theoretical analysis of the influence of the structural parameters on the MIR performance of novel anti-resonant hollow-core fibers. Interestingly, a resonant-like coupling between core modes and modes localized in the glass capillary walls is predicted to occur in the long-wavelength edge of the fundamental transmission band [32], which might have practical implications on the generation and guidance of MIR light. In this regard, chalcogenide glasses are particularly attractive as they present a very broad MIR transmission window extended up to 18 μm, thereby covering the so-called "molecular fingerprinting" region, of crucial importance in spectroscopic and defense applications. Therefore, the paper from NTU Singapore is perfectly complemented by the original research work presented by the *University of Rennes* and *the University of Aix-Marseille* (France). In their joint paper, they present an *Investigation on Chalcogenide Glass Additive Manufacturing for Shaping Mid-Infrared Optical Components and Microstructured Optical Fibers* [33]. An initial step, presented here, is to identify a glass that is suitable for 3D printing. The preform of anti-resonant hollow-core fibers is then printed and subsequently drawn into a fiber. An important aspect presented in this work is that the printing process does not seem to affect the properties of the glass itself. Although imperfections in the printing procedure can however yield additional losses, the presented work certainly shows the potential of the approach, which would greatly ease the fabrication of complex micro-structures.

Specialty PCFs are excellent vehicles for the spectral broadening of a pump laser by exploiting the delicate interplay of third-order nonlinear effects with a tailored dispersion landscape. More recently, the investigation of the polarization properties of these broadband sources [34] has attracted significant attention because of their potential use in spectroscopy [35]. As it is often the case in nonlinear optics, dispersion management holds the key to efficiency, and even gas-filled hollow-core PCFs can yield extremely broad supercontinua [36,37] despite their weak nonlinear material response. In particular, by filling a broadband-guiding hollow-core PCF with noble gases and pumping it with near-infrared ultrashort pulses, it is possible to generate tunable UV radiation via soliton self-compression [38] and resonant dispersive wave emission [39,40]. The original research paper *Photoionization-Induced Broadband Dispersive Wave Generated in an Ar-Filled Hollow-Core Photonic Crystal Fiber* [41] by the *State Key Laboratory of High-Field Laser Physics* and *CAS Center for Excellence in Ultra-Intense Laser Science*, the *Center of Materials Science and Optoelectronics Engineering*, and the *R&D Center of High Power Laser Components* (China) reports on the experimental observation of the spectral broadening of multi-peaked UV dispersive waves via plasma-induced soliton blue-shifting. When the intensities attained upon temporal self-compression of the pump pulses are high enough to cause partial strong-field ionization of the gaseous core, the resulting free-electron cloud strongly modifies the dispersion landscape and affects the nonlinear propagation dynamics [42,43]. This effect is enhanced when the UV emission matches the high-loss bands of the fiber [44].

By contrast with the pressurization of a hollow-core PCF with gas to adjust the dispersion, filling the air channels of a standard solid-core PCF guiding light by modified total internal reflection can drastically affect its guidance mechanism. For example, an endlessly single-mode solid-core PCF can then be turned into a photonic bandgap fiber [45]. In their research paper *Understanding Nonlinear Pulse Propagation in Liquid Strand-Based Photonic Bandgap Fibers* [46], a team from the *University of Jena* (Germany) explores the effects of injecting liquid carbon disulfide (CS_2) in the air channels of a silica-made PCF on the guidance and nonlinear properties of the resulting hybrid fiber, which exhibits distinct transmission bands as previously shown in photonic bandgap fibers [47]. They then generate broad supercontinua in the CS_2-filled fiber, reporting a strong influence of the location of the pump wavelength with respect to the band edge on the dynamics.

Another application of the third-order nonlinearity is the development of sources based on four-wave mixing yielding discrete sidebands, with applications in, e.g., quantum optics. The original research paper *Polarization modulation instability in dispersion-engineered photonic crystal fibers* [48] by the *Universidad de Valencia* (Spain) reports the use of various liquids to fill the air channels of a solid-core PCF to achieve specific conditions, such as an optimal dispersion landscape, for the observation of sidebands created through polarization modulation instability. The thorough experimental demonstration of the phenomenon is supplemented by a rigorous theoretical description providing a solid understanding of the results.

In summary, this special issue showcases the widespread interest that specialty PCF technology still sparks 30 years after its inception. Owing to their current level of maturity and multidisciplinary nature, specialty PCFs are expected to play a key role in multiple scientific, industrial and societal advances in the years to come.

Author Contributions: D.N. and N.Y.J. contributed to the preparation of this manuscript. All authors have read and agreed to the published version of the manuscript.

Funding: This research received no external funding.

Conflicts of Interest: The authors declare no conflict of interest.

References

1. Franken, P.A.; Hill, A.E.; Peters, C.W.; Weinreich, G. Generation of Optical Harmonics. *Phys. Rev. Lett.* **1961**, *7*, 118. [CrossRef]
2. Russell, P. Photonic-Crystal Fibers. *J. Light. Technol.* **2006**, *24*, 4729–4749. [CrossRef]
3. Knight, J.C.; Birks, T.A.; Russell, P.; Atkin, D.M. All-silica single-mode optical fiber with photonic crystal cladding. *Opt. Lett.* **1996**, *21*, 1547–1549. [CrossRef] [PubMed]
4. Russell, P. Photonic Crystal Fibers. *Science* **2003**, *299*, 358–362. [CrossRef] [PubMed]
5. Knight, J.C. Photonic crystal fibres. *Nat. Cell Biol.* **2003**, *424*, 847–851. [CrossRef] [PubMed]
6. Dudley, J.M.; Genty, G.; Coen, S. Supercontinuum generation in photonic crystal fiber. *Rev. Mod. Phys.* **2006**, *78*, 1135–1184. [CrossRef]
7. Villatoro, J.; Kreuzer, M.P.; Jha, R.; Minkovich, V.P.; Finazzi, V.; Badenes, G.; Pruneri, V. Photonic crystal fiber interferometer for chemical vapor detection with high sensitivity. *Opt. Express* **2009**, *17*, 1447–1453. [CrossRef]
8. Michieletto, M.; Lyngs∅, J.C.; Laegsgaard, J.; Bang, O.; Alkeskjold, T.T. Hollow-core fibers for high power pulse delivery. *Opt. Express* **2016**, *24*, 7103–7119. [CrossRef]
9. Balciunas, T.; Dutin, C.F.; Fan, G.; Witting, T.; Voronin, A.; Zheltikov, A.; Gérôme, F.; Paulus, G.G.; Baltuska, A.; Benabid, F.A. A strong-field driver in the single-cycle regime based on self-compression in a kagome fibre. *Nat. Commun.* **2015**, *6*, 6117. [CrossRef]
10. Poletti, F.; Wheeler, N.V.; Petrovich, M.N.; Baddela, N.; Fokoua, E.R.N.; Hayes, J.R.; Gray, D.R.; Li, Z.; Slavik, R.; Richardson, D.; et al. Towards high-capacity fibre-optic communications at the speed of light in vacuum. *Nat. Photonics* **2013**, *7*, 279–284. [CrossRef]
11. Finger, M.A.; Iskhakov, T.S.; Joly, N.Y.; Chekhova, M.; Russell, P. Raman-Free, Noble-Gas-Filled Photonic-Crystal Fiber Source for Ultrafast, Very Bright Twin-Beam Squeezed Vacuum. *Phys. Rev. Lett.* **2015**, *115*, 143602. [CrossRef]
12. Rifat, A.A.; Ahmed, R.; Yetisen, A.K.; Butt, H.; Sabouri, A.; Amouzad Mahdiraji, G.; Yun, S.H.; Mahamd Adikan, F.R. *Sensors and Actuators B: Chemical*; Elsevier: Amsterdam, The Netherlands, 2017; Volume 243, pp. 311–325.
13. Koeppel, M.; Sharma, A.; Podschus, J.; Sundaramahalingam, S.; Joly, N.Y.; Xie, S.; Russell, P.S.J.; Schmauss, B. Doppler optical frequency domain reflectometry for remote fiber sensing. *Opt. Express* **2021**, *29*, 14615–14629. [CrossRef]
14. Abdolghader, P.; Pegoraro, A.F.; Joly, N.Y.; Ridsdale, A.; Lausten, R.; Légaré, F.; Stolow, A. All normal dispersion nonlinear fibre supercontinuum source characterization and application in hyperspectral stimulated Raman scattering microscopy. *Opt. Express* **2020**, *28*, 35997–36008. [CrossRef] [PubMed]
15. Jasion, G.T.; Bradley, T.D.; Harrington, K.; Sakr, H.; Chen, Y.; Fokoua, E.N.; Davidson, I.A.; Taranta, A.; Hayes, J.R.; Richardson, D.J.; et al. Hollow Core NANF with 0.28 dB/km Attenuation in the C and L Bands. In Proceedings of the Optical Fiber Communication Conference Postdeadline Papers 2020, San Diego, CA, USA, 8–12 March 2020; p. 4.
16. Gebert, F.; Frosz, M.H.; Weiss, T.; Wan, Y.; Ermolov, A.; Joly, N.Y.; Schmidt, P.O.; Russell, P.S.J. Damage-free single-mode transmission of deep-UV light in hollow-core PCF. *Opt. Express* **2014**, *22*, 15388–15396. [CrossRef] [PubMed]
17. Jiang, X.; Joly, N.Y.; Finger, M.A.; Babic, F.; Wong, G.; Travers, J.; Russell, P.S.J. Deep-ultraviolet to mid-infrared supercontinuum generated in solid-core ZBLAN photonic crystal fibre. *Nat. Photonics* **2015**, *9*, 133–139. [CrossRef]
18. Beravat, R.; Wong, G.K.L.; Frosz, M.H.; Xi, X.M.; Russell, P.S. Twist-induced guidance in coreless photonic crystal fiber: A helical channel for light. *Sci. Adv.* **2016**, *2*, e1601421. [CrossRef] [PubMed]
19. Davtyan, S.; Novoa, D.; Chen, Y.; Frosz, M.H.; Russell, P.S.J. Polarization-Tailored Raman Frequency Conversion in Chiral Gas-Filled Hollow-Core Photonic Crystal Fibers. *Phys. Rev. Lett.* **2019**, *122*, 143902. [CrossRef] [PubMed]
20. Limpert, J.; Deguil-Robin, N.; Manek-Hönninger, I.; Salin, F.; Röser, F.; Liem, A.; Schreiber, T.; Nolte, S.; Zellmer, H.; Tünnermann, A.; et al. High-power rod-type photonic crystal fiber laser. *Opt. Express* **2005**, *13*, 1055–1058. [CrossRef] [PubMed]
21. Cubillas, A.M.; Unterkofler, S.; Euser, T.; Etzold, B.J.M.; Jones, A.C.; Sadler, P.J.; Wasserscheid, P.; Russell, P.S. Photonic crystal fibres for chemical sensing and photochemistry. *Chem. Soc. Rev.* **2013**, *42*, 8629–8648. [CrossRef]

22. Schorn, F.; Aubermann, M.; Zeltner, R.; Haumann, M.; Joly, N.Y. Online Monitoring of Microscale Liquid-Phase Catalysis Using in-Fiber Raman Spectroscopy. *ACS Catal.* **2021**, *11*, 6709–6714. [CrossRef]
23. Li, J.; Li, H.; Wang, Z. Application of Hollow-Core Photonic Crystal Fibers in Gas Raman Lasers Operating at 1.7 μm. *Crystals* **2021**, *11*, 121. [CrossRef]
24. Mridha, M.K.; Novoa, D.; Bauerschmidt, S.T.; Abdolvand, A.; Russell, P. Generation of a vacuum ultraviolet to visible Raman frequency comb in H_2-filled kagomé photonic crystal fiber. *Opt. Lett.* **2016**, *41*, 2811–2814. [CrossRef]
25. Tyumenev, R.; Russell, P.S.J.; Novoa, D. Narrowband Vacuum Ultraviolet Light via Cooperative Raman Scattering in Dual-Pumped Gas-Filled Photonic Crystal Fiber. *ACS Photonics* **2020**, *7*, 1989–1993. [CrossRef]
26. Couny, F.; Benabid, F.; Roberts, P.J.; Light, P.S.; Raymer, M.G. Generation and Photonic Guidance of Multi-Octave Optical-Frequency Combs. *Science* **2007**, *318*, 1118–1121. [CrossRef] [PubMed]
27. Hosseini, P.; Novoa, D.; Abdolvand, A.; Russell, P.S. Enhanced Control of Transient Raman Scattering Using Buffered Hydrogen in Hollow-Core Photonic Crystal Fibers. *Phys. Rev. Lett.* **2017**, *119*, 253903. [CrossRef] [PubMed]
28. Benabid, F.; Bouwmans, G.; Knight, J.; Russell, P.S.J.; Couny, F. Ultrahigh Efficiency Laser Wavelength Conversion in a Gas-Filled Hollow Core Photonic Crystal Fiber by Pure Stimulated Rotational Raman Scattering in Molecular Hydrogen. *Phys. Rev. Lett.* **2004**, *93*, 123903. [CrossRef] [PubMed]
29. Gladyshev, A.; Yatsenko, Y.; Kolyadin, A.; Kompanets, V.; Bufetov, I. Mid-infrared 10-μJ-level sub-picosecond pulse generation via stimulated Raman scattering in a gas-filled revolver fiber. *Opt. Mater. Express* **2020**, *10*, 3081–3089. [CrossRef]
30. Pei, W.; Li, H.; Huang, W.; Wang, M.; Wang, Z. Hydrogen Molecules Rotational Stimulated Raman Scattering in All-Fiber Cavity Based on Hollow-Core Photonic Crystal Fibers. *Crystals* **2021**, *11*, 711. [CrossRef]
31. Deng, A.; Chang, W. Geometrical Scaling of Antiresonant Hollow-Core Fibers for Mid-Infrared Beam Delivery. *Crystals* **2021**, *11*, 420. [CrossRef]
32. Cassataro, M.; Novoa, D.; Günendi, M.C.; Edavalath, N.N.; Frosz, M.H.; Travers, J.; Russell, P. Generation of broadband mid-IR and UV light in gas-filled single-ring hollow-core PCF. *Opt. Express* **2017**, *25*, 7637–7644. [CrossRef]
33. Carcreff, J.; Cheviré, F.; Lebullenger, R.; Gautier, A.; Chahal, R.; Adam, J.; Calvez, L.; Brilland, L.; Galdo, E.; Le Coq, D.; et al. Investigation on Chalcogenide Glass Additive Manufacturing for Shaping Mid-infrared Optical Components and Microstructured Optical Fibers. *Crystals* **2021**, *11*, 228. [CrossRef]
34. Sopalla, R.P.; Wong, G.K.L.; Joly, N.Y.; Frosz, M.H.; Jiang, X.; Ahmed, G.; Russell, P.S. Generation of broadband circularly polarized supercontinuum light in twisted photonic crystal fibers. *Opt. Lett.* **2019**, *44*, 3964–3967. [CrossRef]
35. Couture, N.; Ostic, R.; Reddy, P.H.; Kar, A.K.; Paul, M.C.; Ménard, J.-M. Polarization-resolved supercontinuum generated in a germania-doped photonic crystal fiber. *J. Phys. Photonics* **2021**, *3*, 025002. [CrossRef]
36. Belli, F.; Abdolvand, A.; Chang, W.; Travers, J.; Russell, P. Vacuum-ultraviolet to infrared supercontinuum in hydrogen-filled photonic crystal fiber. *Optica* **2015**, *2*, 292–300. [CrossRef]
37. Elu, U.; Maidment, L.; Vamos, L.; Tani, F.; Novoa, D.; Frosz, M.H.; Badikov, V.; Badikov, D.; Petrov, V.; Russell, P.S.J.; et al. Seven-octave high-brightness and carrier-envelope-phase-stable light source. *Nat. Photonics* **2021**, *15*, 277–280. [CrossRef]
38. Russell, P.; Hoelzer, P.; Chang, W.; Abdolvand, A.; Travers, J. Hollow-core photonic crystal fibres for gas-based nonlinear optics. *Nat. Photonics* **2014**, *8*, 278–286. [CrossRef]
39. Joly, N.Y.; Nold, J.; Chang, W.; Hölzer, P.; Nazarkin, A.; Wong, G.; Biancalana, F.; Russell, P. Bright Spatially Coherent Wavelength-Tunable Deep-UV Laser Source Using an Ar-Filled Photonic Crystal Fiber. *Phys. Rev. Lett.* **2011**, *106*, 203901. [CrossRef]
40. Mak, K.F.; Travers, J.; Hölzer, P.; Joly, N.Y.; Russell, P. Tunable vacuum-UV to visible ultrafast pulse source based on gas-filled Kagome-PCF. *Opt. Express* **2013**, *21*, 10942–10953. [CrossRef] [PubMed]
41. Fu, J.; Chen, Y.; Huang, Z.; Yu, F.; Wu, D.; Pan, J.; Zhang, C.; Wang, D.; Pang, M.; Leng, Y. Photoionization-Induced Broadband Dispersive Wave Generated in an Ar-Filled Hollow-Core Photonic Crystal Fiber. *Crystals* **2021**, *11*, 180. [CrossRef]
42. Novoa, D.; Cassataro, M.; Travers, J.; Russell, P. Photoionization-Induced Emission of Tunable Few-Cycle Midinfrared Dispersive Waves in Gas-Filled Hollow-Core Photonic Crystal Fibers. *Phys. Rev. Lett.* **2015**, *115*, 033901. [CrossRef] [PubMed]
43. Köttig, F.; Novoa, D.; Tani, F.; Günendi, M.C.; Cassataro, M.; Travers, J.; Russell, P.S.J. Mid-infrared dispersive wave generation in gas-filled photonic crystal fibre by transient ionization-driven changes in dispersion. *Nat. Commun.* **2017**, *8*, 1–8. [CrossRef] [PubMed]
44. Tani, F.; Köttig, F.; Novoa, D.; Keding, R.; Russell, P.S. Effect of anti-crossings with cladding resonances on ultrafast nonlinear dynamics in gas-filled photonic crystal fibers. *Photonics Res.* **2018**, *6*, 84–88. [CrossRef]
45. Luan, F.; George, A.K.; Hedley, T.D.; Pearce, G.J.; Bird, D.M.; Knight, J.C.; Russell, P. All-solid photonic bandgap fiber. *Opt. Lett.* **2004**, *29*, 2369–2371. [CrossRef] [PubMed]
46. Qi, X.; Schaarschmidt, K.; Li, G.; Junaid, S.; Scheibinger, R.; Lühder, T.; Schmidt, M. Understanding Nonlinear Pulse Propagation in Liquid Strand-Based Photonic Bandgap Fibers. *Crystals* **2021**, *11*, 305. [CrossRef]
47. Argyros, A.; Birks, T.A.; Leon-Saval, S.G.; Cordeiro, C.M.B.; Russell, P.S.J. Guidance properties of low-contrast photonic bandgap fibres. *Opt. Express* **2005**, *13*, 2503–2511. [CrossRef]
48. Loredo-Trejo, A.; Díez, A.; Silvestre, E.; Andrés, M. Polarization Modulation Instability in Dispersion-Engineered Photonic Crystal Fibers. *Crystals* **2021**, *11*, 365. [CrossRef]

 MDPI

Review

Application of Hollow-Core Photonic Crystal Fibers in Gas Raman Lasers Operating at 1.7 μm

Jun Li [1], Hao Li [2] and Zefeng Wang [2,3,*]

[1] Environmental Economy and Information Department, Changsha Environmental Protection Vocational College, Changsha 410004, China; lijun@cshbxy.com
[2] College of Advanced Interdisciplinary Studies, National University of Defense Technology, Changsha 410073, China; lihao18c@nudt.edu.cn
[3] State Key Laboratory of Pulsed Power Laser Technology, Changsha 410073, China
* Correspondence: zefengwang@nudt.edu.cn or zefengwang@nudt.edu.com

Abstract: A 1.7 μm pulsed laser plays an important role in bioimaging, gas detection, and so on. Fiber gas Raman lasers (FGRLs) based on hollow-core photonic crystal fibers (HC-PCFs) provide a novel and effective method for fiber lasers operating at 1.7 μm. Compared with traditional methods, FGRLs have more advantages in generating high-power 1.7 μm pulsed lasers. This paper reviews the studies of 1.7 μm FGRLs, briefly describes the principle and characteristics of HC-PCFs and gas-stimulated Raman scattering (SRS), and systematical characterizes 1.7 μm FGRLs in aspects of output spectral coverage, power-limiting factors, and a theoretical model. When the fiber length and pump power are constant, a relatively high gas pressure and appropriate pump peak power are the key to achieving high-power 1.7 μm Raman output. Furthermore, the development direction of 1.7 μm FGRLs is also explored.

Keywords: stimulated Raman scattering; hollow-core photonic crystal fibers; fiber lasers; gas lasers

Citation: Li, J.; Li, H.; Wang, Z. Application of Hollow-Core Photonic Crystal Fibers in Gas Raman Lasers Operating at 1.7 μm. *Crystals* **2021**, *11*, 121. https://doi.org/10.3390/cryst11020121

Academic Editors: David Novoa and Nicolas Y. Joly

Received: 30 December 2020
Accepted: 25 January 2021
Published: 27 January 2021

1. Introduction

In recent years, lasers operating in the 1.7 μm band (1650–1750 nm) have received much attention due to the growing number of promising applications, such as bioimaging, gas detection, medical treatment, and mid-infrared laser generation. Compared with other types of lasers, 1.7 μm fiber lasers have been more intensively studied due to good stability, compactness, and beam quality. However, reported works mainly involve 1.7 μm continuous-wave (CW) fiber lasers [1,2], and the pulsed fiber lasers operating in this wavelength region have not been researched fully, though they have unique advantages in some applications. For example, in bioimaging applications such as multiphoton microscopy [3,4], optical coherence tomography [5], and spectroscopic photoacoustic (PA) imaging [6,7], 1.7 μm pulsed lasers can be used to realize three-dimensional (3D) volumetric imaging by time-resolved ultrasonic detection [7]. Similarly, in methane detection, the 3D distribution of the CH_4 concentration in space can be measured by using the time-of-flight ranging method to analyze the temporal characteristics of pulsed lasers when the pulsed lasers are in the 1.7 μm band used as the detection signal [8]. In fact, whether in bioimaging or gas detection, high-power 1.7 μm pulsed lasers are conducive to achieving higher sensitivity and deeper penetration/detection [7,9]. Thus, it is important and necessary to increase the output power of 1.7 μm pulsed fiber lasers.

The traditional methods of realizing 1.7 μm pulsed fiber lasers can be mainly classified into two categories. One is based on population inversion to generate 1.7 μm pulsed lasers by using rare-earth-doped fibers, such as thulium-doped fibers [6,7,10,11], thulium–holmium-codoped fibers [12,13], and bismuth-doped fibers [14–16]. The other is based on nonlinear effects in solid-core fibers to realize a frequency conversion, such as soliton self-frequency shift [17–20], four-wave mixing [21–23], self-phase modulation [7,24,25], and stimulated Raman scattering (SRS) [26]. Recently, fiber gas Raman lasers (FGRLs) based on hollow-core photonic crystal fibers (HC-PCFs) have opened a new opportunity for 1.7 μm pulsed fiber lasers [27–30]. The principle of FGRLs is to realize a frequency conversion by gas SRS in HC-PCFs. The output average powers and corresponding pulse widths of reported 1.7 μm pulsed fiber lasers are plotted in Figure 1.

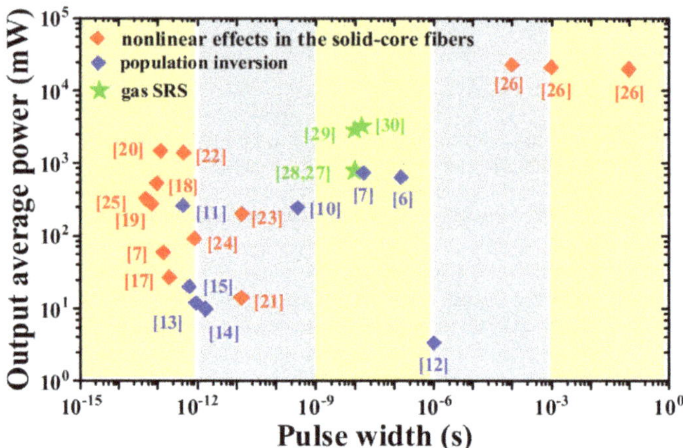

Figure 1. The output average powers and corresponding pulse widths of reported 1.7 μm pulsed fiber lasers.

It can be seen that the pulse widths of 1.7 μm pulsed lasers based on solid-core fibers are mostly in femtosecond and picosecond regions and their output average powers are mostly less than 1 W. Only a pulsed Raman fiber laser can achieve an output power of up to 23 W [26], but its pulse width is at the millisecond level. Thus, it is very challenging to obtain high-power 1.7 μm pulsed lasers with a short pulse width based on solid-core fibers. The 1.7 μm FGRLs fill this gap and achieve nanosecond pulsed lasers with an output average power of over 3 W [30]. In fact, 1.7 μm nanosecond short pulsed lasers have more advantages in bioimaging. To realize volumetric imaging with higher resolution, the pulse duration should be around 10 ns or even shorter [7]. Furthermore, 1.7 μm FGRLs with continuous wavelength tunability and narrow linewidth also have more advantages in gas detection. Continuous wavelength tunability is conducive to the detection of different kinds of gas molecules, and narrow-linewidth pulsed lasers can not only accurately distinguish the absorption peaks but also have a longer coherent distance that is helpful in long-distance detection [9]. Therefore, compared with traditional 1.7 μm pulsed fiber lasers, 1.7 μm FGRLs have unique advantages and strong competitiveness due to the characteristics of high power, high efficiency, continuous wavelength tunability, and a narrow linewidth.

In this paper, we review the studies of 1.7 μm FGRLs based on HC-PCFs. Section 2 briefly introduces the classification, light guide mechanism, and development of HC-PCFs. In Section 3, a comparison of gas SRS in free space and in HC-PCFs is made and the characteristics of candidate Raman gas media for 1.7 μm FGRLs are discussed. Section 4 describes the typical experimental structure of 1.7 μm FGRLs and characterizes 1.7 μm FGRLs in aspects of output spectral coverage, power-limiting factors, and a theoretical model. When the fiber length and pump power are constant, a relatively high gas pressure and appropriate pump peak power are the key to achieving high output Raman power in 1.7 μm FGRLs. In Section 5, we discuss the future development of 1.7 μm FGRLs based on HC-PCFs.

2. Hollow-Core Photonic Crystal Fibers

HC-PCFs are the core components of FGRLs, which provide a platform for the interaction of laser and gas. The fiber core of an HC-PCF is an air hole, so the refractive index of the fiber core is less than that of the fiber cladding and the law of total reflection is not suitable for HC-PCFs. According to the light guide mechanism, there are two major classifications of HC-PCFs, namely photonic bandgap hollow-core fibers (PBG-HCFs) [31–35] and anti-resonance hollow-core fibers (AR-HCFs) (or inhibited-coupling fibers) [36–49].

2.1. *Photonic Bandgap Hollow-Core Fibers*

In 1999, Russell et al. demonstrated the first PBG-HCFs [31], the schematic cross section of which is shown in Figure 2a. It can be seen that the fiber core is a larger air hole and periodic small air holes are distributed in the cladding of the PBG-HCFs. These periodic air holes in the cladding form a two-dimensional photonic bandgap, so light cannot pass through the cladding and be confined in the fiber core when the wavelength of light is located in the photonic bandgap. Since then, HC-PCFs have been developed toward lower loss and greater bandwidths. In 2004, the University of Bath demonstrated low-loss PBG-HCFs of 1.72 dB/km at 1565 nm [32], the schematic cross section of which is shown in Figure 2b. In 2005, they further reduced the fiber loss to 1.2 dB/km at 1620 nm [33], which is currently the lowest loss of PBG-HCFs. Figure 2c shows the schematic cross section of PBG-HCFs with a broad transmission band and low loss demonstrated by the Beijing University of Technology in 2019 [34]. The minimum loss of 6.5 dB/km at 1633 nm and a 3 dB bandwidth at 458 nm were achieved, which is the broadest bandwidth in PBG-HCFs. The main factors affecting the loss of PBG-HCFs are the scattering caused by the surface roughness of the fiber core boundary [33] and the coupling between the core and the surface modes [35]. Moreover, the coupling also causes multiple loss peaks in the transmission band, affecting the transmission bandwidth of the PBG-HCFs.

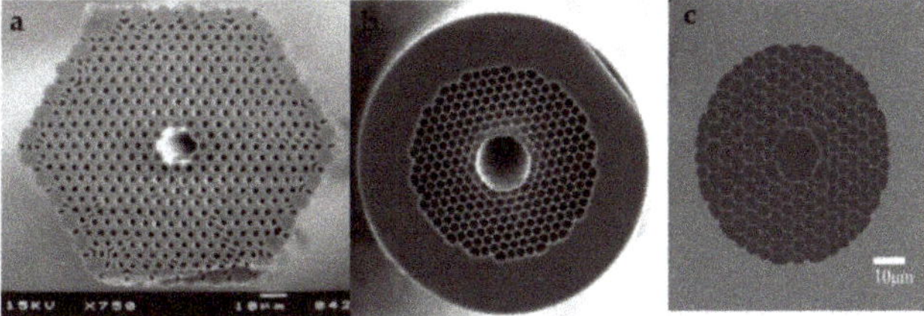

Figure 2. The schematic cross section of HC-PCFs. (**a**) The first PBG-HCF [31], (**b**) low-loss PBG-HCF [32], and (**c**) broadband PBG-HCF [34].

2.2. *Anti-Resonance Hollow-Core Fibers*

The Kagome HCFs reported in 2002 are the first AR-HCFs [36], the schematic cross section of which is shown in Figure 3a. Compared with PBG-HCFs, the cladding structure pitch of Kagome HCFs is larger, which can widen the optical transmission band [37]. Furthermore, there is no complete photonic bandgap in the Kagome hollow fiber, and its light-guiding mechanism can be explained by inhibited coupling [38] or the anti-resonant reflection optical waveguide (ARROW) [39]. According to the inhibited-coupling mechanism, a core-guide mode can be strongly inhibited from channeling out through the cladding by a mismatch between the core and cladding modes [40]. According to ARROW, the microstructure in the cladding is similar to a Fabry–Perot cavity. The light that meets the cavity resonance conditions will leak out through the cladding, while the light that cannot resonate in the cavity is prevented from leaking from the cladding and confined to the fiber core [41]. Figure 3b shows the schematic cross section of hypocycloid Kagome HCFs, and the design of the fiber core boundary with negative curvature reduces the transmission loss of the fiber [42]. With further study of the light-guiding mechanism of AR-HCFs, it has been found that the transmission performance of AR-HCFs is mainly determined by the first ring microstructure of the fiber core boundary. Subsequently, AR-HCFs with simpler microstructures and better optical performance have emerged [43–49], and all of these AR-HCFs obey the ARROW guiding mechanism. Figure 2c presents the schematic cross section of the first tube-structure AR-HCFs reported in 2011, the cladding of which is composed of single-ring tubes [44]. The attenuation of the single-ring AR-HCFs in the mid-infrared band is much lower than that of the silica glass solid-core fiber. Figure 3d shows the schematic cross section of ice-cream-type AR-HCFs [45], the transmission band of which is located in the mid-infrared band and the minimum loss of which is 34 dB/km at 3050 nm. With further study of AR-HCFs, it has been found that the touching points of adjacent capillaries in the cladding behave as independent waveguides supporting their own lossy modes, which would introduce additional transmission loss [50,51]. Thus, nodeless single-ring AR-HCFs were first demonstrated in 2013 [46]. In addition, it was demonstrated that the elimination of the tube's contact point helps to reduce the bending loss [47]. In 2019, nodeless nested AR-HCFs with the attenuation of 0.65 dB/km in the C and L telecommunications bands were demonstrated, which means this was the first time that HC-PCFs realized a loss comparable to that of silica glass solid-core fibers [48]. The attenuation of nodeless nested AR-HCFs was further recued to 0.28 dB/km in 2020 [49].

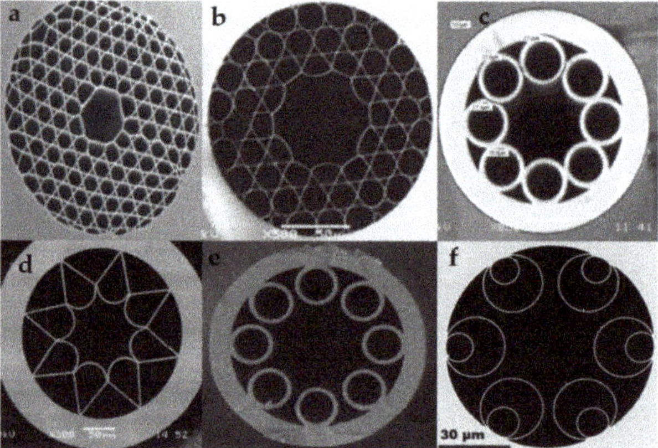

Figure 3. The schematic cross section of HC-PCFs. (**a**) The first AR-HCF [36], (**b**) hypocycloid Kagome AR-HCFs [42], (**c**) single-ring AR-HCFs [44], (**d**) ice-cream-type AR-HCF [45], (**e**) nodeless single-ring AR-HCFs [46], and (**f**) nodeless nested AR-HCFs [48].

Compared with PBG-HCFs, the mode field of AR-HCFs overlaps the silica glass of the fiber core boundary less [52]. Thus, the surface scattering loss of AR-HCFs is lower than that of PBG-HCFs, which means that AR-HCFs have advantages of realizing lower loss and reducing the nonlinearities caused by silica glass. Furthermore, although both AR-HCFs and PBG-HCFs are multimode, the attenuation of high-order modes in AR-HCFs is higher than that of PBG-HCFs, which means that with AR-HCFs it is easier to obtain a fundamental mode guidance at a short fiber length [53].

3. Gas Stimulated Raman Scattering

3.1. Traditional Gas Raman Lasers Versus Fiber Gas Raman Lasers

Since gas SRS was first reported in 1963 [54], gas SRS has been considered as a significant method to realize a laser of a new wavelength by frequency conversion. Gas SRS can cover the ultraviolet-to-infrared band [55,56], which is an effective extension of the laser output band. Figure 4a is a simplified schematic diagram of the interaction between lasers and gas in traditional Raman gas lasers. Owing to the diffraction effect of laser transmission in free space and some other factors, the gas SRS in free space has a short interaction length and low intensity, leading to an extremely high Raman threshold. Normally, traditional gas Raman lasers require pump pulsed lasers with megawatt peak power to reach the Raman threshold. Moreover, there would be many Stokes waves generated, so it is difficult to realize efficient conversion of a single Stokes wavelength in traditional gas Raman lasers. These problems limit the applications of traditional gas Raman lasers. However, for FGRLs, gas SRS occurs in HC-PCFs, and a simplified schematic diagram of the interaction between lasers and gas in FGRLs is shown in Figure 4b. HC-PCFs can confine lasers to the small core for long-distance transmission, so an extremely low Raman threshold can be achieved. Furthermore, the output Stokes waves can be controlled by designing the transmission band of HC-PCFs. Thus, HC-PCFs provide an ideal environment for efficient gas SRS, meeting the requirements of strong interaction intensity, a long interaction length, and a controllable Raman gain spectrum at the same time.

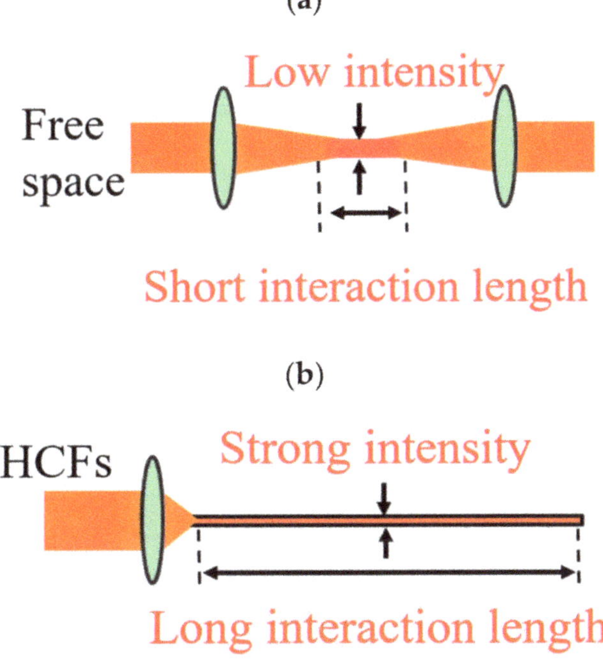

Figure 4. The simplified schematic diagram of the interaction between lasers and gas in (**a**) traditional gas Raman lasers and (**b**) fiber gas Raman lasers.

3.2. Candidate Gas Media for 1.7 μm Raman Wavelength

For gas SRS, the Raman frequency shift is due to the energy-level transitions of gas media. Thus, different output Raman wavelengths can be flexibly realized by changing the gas media or using different energy-level transitions of one gas medium. Because hydrogen (H_2) and deuterium (D_2) are the main candidate gas media for 1.7 μm FGRLs, Figure 5a,b present the schematic diagrams of the energy-level transitions of H_2 [57] and D_2 SRS [58], respectively. It can be seen that H_2 molecules' vibrational SRS has a Raman frequency shift coefficient of ~4155 cm^{-1} and the rotational SRS of different energy levels has Raman frequency shift coefficients of ~587, ~354, and ~814 cm^{-1}. D_2 molecules' vibrational SRS has a Raman frequency shift coefficient of ~2988 cm^{-1}, and the rotational SRS of different energy levels has Raman frequency shift coefficients of ~179, ~297, and ~415 cm^{-1}.

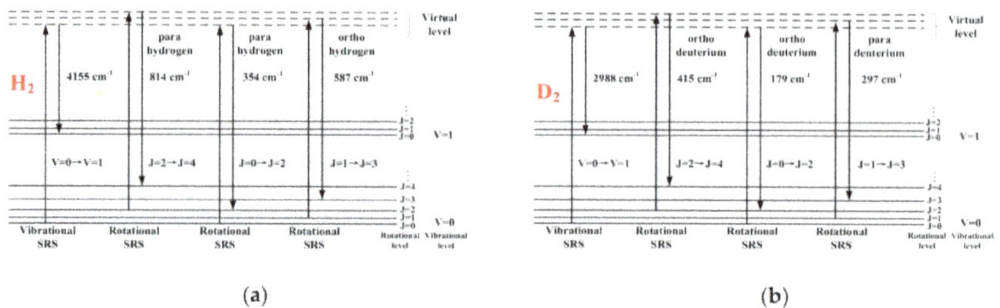

Figure 5. The schematic diagram of the energy-level transitions of (**a**) H_2 [57] and (**b**) D_2 [58] SRS.

The relationship between the pump wavelength, Raman frequency shift coefficient, and Raman wavelength is given by

$$\Delta\omega = \frac{1}{\lambda_p} - \frac{1}{\lambda_s} \tag{1}$$

where $\Delta\omega$ is the Raman frequency shift coefficient and λ_P and λ_S are the pump wavelength and Raman wavelength, respectively. When the pump wavelength is set in the 1 or 1.5 μm band, H_2 or D_2 SRS can realize different output Raman wavelengths, as shown in Table 1. It can be seen that when the pump wavelength is set in the 1.5 μm band, the output Raman wavelength at the 1.7 μm band can be realized by H_2 or D_2 rotational SRS.

Table 1. Different pump wavelengths, Raman frequency shift, and corresponding output Raman wavelengths.

Gain Gases	Raman Frequency Shift	Raman Wavelength Pumped at 1064 nm	Raman Wavelength Pumped at 1550 nm
H_2	4155 cm^{-1}	1907 nm	4354 nm
	814 cm−1	1165 nm	1773 nm
	587 cm−1	1135 nm	1705 nm
	354 cm^{-1}	1106 nm	1640 nm
D_2	2987 cm^{-1}	1560 nm	2886 nm
	415 cm^{-1}	1113 nm	1656 nm
	297 cm^{-1}	1098 nm	1625 nm
	179 cm^{-1}	1084 nm	1594 nm

4. Fiber Gas Raman Lasers Operating at 1.7 μm

4.1. Typical Experimental Setup

Figure 6a presents the typical experimental setup of 1.7 μm FGRLs, comprising mainly a pump source, HC-PCFs, a gas cell, and a set of lenses. The pump laser is a 1.5 μm pulsed erbium-doped fiber amplifier (EDFA), the pigtail fiber (Corning, SMF-28e) of which is fusion-spliced with the HC-PCFs (NKT Photonics, HC-1550-02) due to a similar mode field area and numerical aperture, and the theoretical minimum loss can be as low as 1.3 dB [59]. The output end of the HC-PCFs is sealed in a gas cell with a glass window, and the HC-PCFs can be filled with a gas medium through the gas cell. The output lasers being transmitted through the glass window are collimated and are filtered by a set of lenses.

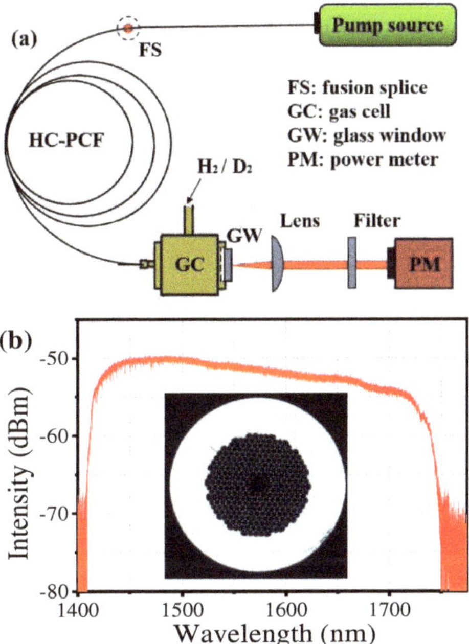

Figure 6. (**a**) The schematic diagram of a typical experimental setup of 1.7 μm FGRLs and (**b**) the transmission spectrum of the used HC-PCFs. Insert: optical micrograph of a cross section of the used HC-PCFs [57].

Figure 6b presents the transmission spectrum and optical micrograph of a cross section of HC-PCFs [57]. The low-loss transmission band of HC-PCFs is from 1415 to 1740 nm, covering both the 1.5 μm pump wavelengths and the 1.7 μm output Raman wavelengths. Furthermore, it can be seen from the insert that the used HC-PCFs are PBG-HCFs with a multilayer microstructure in the cladding. Compared with the AR-HCFs with a single-ring microstructure in the cladding, fusion splicing has less impact on the optical performance of the used HC-PCFs. Thus, mature commercial HC-PCFs (NKT Photonics, HC-1550-02) are very suitable for 1.7 μm FGRLs.

4.2. Spectral Coverage

A pulsed fiber amplifier with a wavelength tuning range of 1535–1565 nm was used to pump 20-m-long HC-PCFs filled with D_2 or H_2, and the output spectra are shown in Figure 7a [27] and Figure 7b [28], respectively. It can be seen that one pump line in the 1.5 um band is converted into one Raman line in the 1.7 μm band (415 and 587 cm^{-1} Raman frequency shift coefficients of D_2 and H_2, respectively), which greatly improves the power conversion efficiency. This is not difficult to explain. Take D_2 SRS as an example; if the pump wavelength is 1550 nm, because the 2886 nm Raman line generated by vibrational SRS (with a Raman frequency shift of 2987 cm^{-1}) is located outside the low-loss transmission band, it is strongly suppressed. Moreover, because the Raman gain of the 1656 nm rotational Raman line (with a Raman frequency shift of 415 cm^{-1}) is higher than that of 1625 and 1594 nm rotational Raman lines (with Raman frequency shifts of 297 and 179 cm^{-1}, respectively), the 1550 nm pump line will first be converted into a 1656 nm Raman line, and the residual pump power is too low to generate 1625 and 1594 nm Raman lines. The explanation of the output spectrum of H_2 SRS is also similar. Therefore, the output Raman wavelengths are determined by the gas media, pump wavelength, and fiber attenuation.

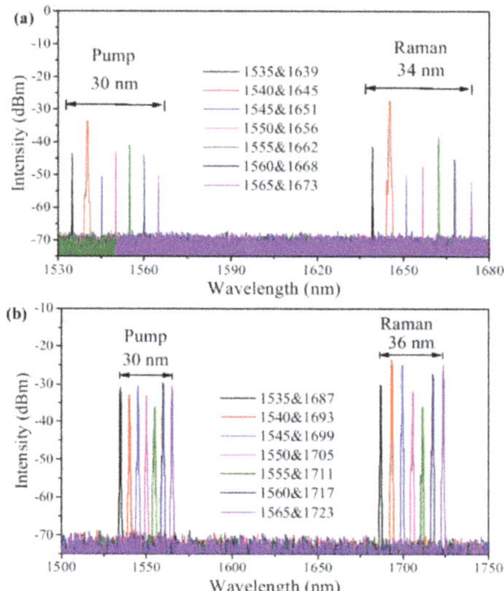

Figure 7. The output spectra of 1.7 μm FGRLs based on (**a**) D_2-filled [27] and (**b**) H_2-filled HC-PCFs [28] when the pump wavelength is from 1535 to 1565 nm and the fiber length is 20 m.

4.3. Power-Limiting Factors

Compared to the solid-core fiber, only less than 1% of the mode field energy overlaps the silica glass of the fiber core boundary when the lasers are transmitted in HC-PCFs. So HC-PCFs have a higher damage threshold, which means that HC-PCFs are more capable of transmitting comparatively higher power and energy. In this section, the influence of the pump wavelength, gas pressure, repetition frequency of the pump pulse, and fiber length on the output Raman power is fully discussed based on the reported experimental results of the 1.7 μm fiber deuterium gas Raman laser [27,29], and the key factors for achieving high-power 1.7 μm FGRLs are analyzed.

Figure 8a shows the output Raman power as a function of coupled average pump power at different pump wavelengths when the fiber length is 20 m [27]. The maximum Raman power decreases toward the long wavelength. It can be attributed to the amplification performance of the pump source. The output power of the pump source is slightly reduced as the wavelength increases. In fact, when the fiber attenuation in the pump band and output Raman band is basically unchanged, the wavelength change has little effect on the output Raman power. Figure 8b shows the output Raman power as a function of coupled average pump power at different gas pressures when the fiber length is 20 m [27]. Since the molecular density of the gain gas in the HC-PCFs can be adjusted by the barometer of the gas cell, this provides new freedom in the optimization of the 1.7 μm FGRL performance for high-power laser emission. When the gas pressure is too low, the Raman threshold is relatively higher due to the small Raman gain [60], thereby affecting the output Raman power. While the Raman gain is saturated at a high gas pressure level, the contribution of increasing gas pressure to the Raman gain becomes extremely small. Therefore, it can be seen from Figure 7b that when the gas pressure is higher than 15 bar, the Raman power does not increase obviously.

Figure 9 presents the experimental results of using a higher-power pump source [29], and cascaded Raman conversion is more likely to occur under high pump power. Figure 9a presents the output Raman power as a function of the coupled average pump power at different repetition frequencies of the pump pulse when the fiber length is 20 m and the gas pressure is 16 bar. It can be seen that the repetition frequency has a great influence on the output Raman power, and there is an optimal repetition frequency to achieve the maximum output Raman power. This is because when the fiber length and the gas pressure are constant, the Raman threshold of the pump peak power is also constant. Thus, it is necessary to adjust the repetition frequency to obtain a suitable peak power so that the peak power is much higher than the first-order Raman threshold but does not exceed the second-order Raman threshold. When the peak power is much higher than the first-order Raman threshold, more pump pulse energy can undergo Raman conversion. When the peak power does not exceed the second-order Raman threshold, the first-order to second-order cascaded Raman conversion will not occur, which means that the first-order Raman power will not fall at the high pump power level. Figure 8b presents the output Raman power as a function of the coupled average pump power at different fiber lengths when the repetition is 1.5 MHz and the gas pressure is 16 bar. It can be seen that the Raman threshold increases with a decrease in the fiber length, which means that less pump pulse energy is converted into Raman pulse energy, so the maximum output Raman power also decreases. Therefore, when the fiber length is reduced, a higher peak power of the pump pulse is required to improve the power conversion efficiency.

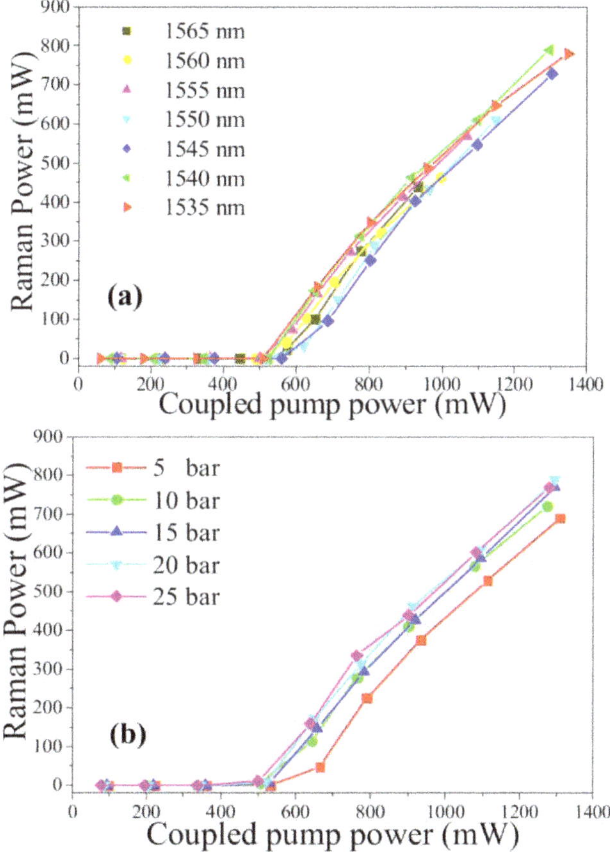

Figure 8. The output Raman power as a function of coupled pump power at different (**a**) pump wavelengths and (**b**) gas pressures when the fiber length is 20 m [27].

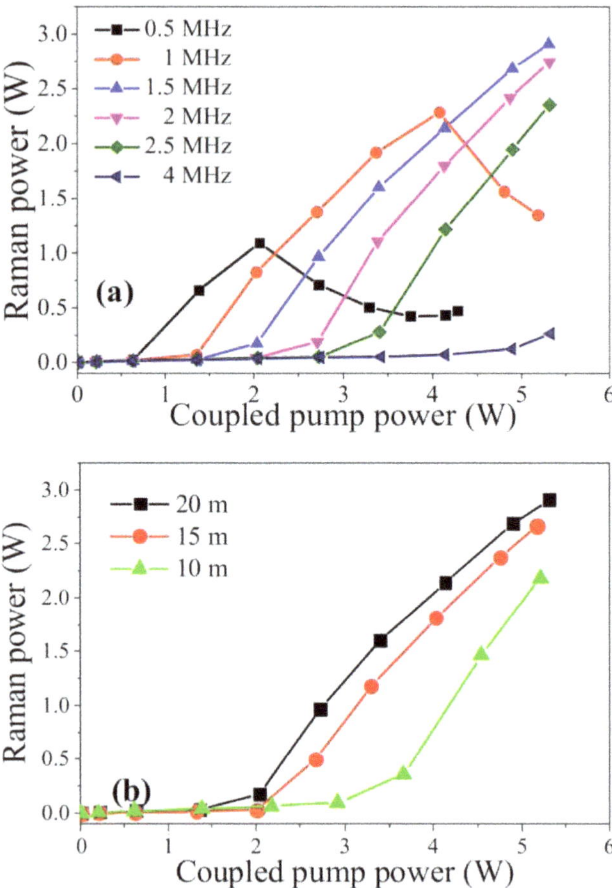

Figure 9. The output Raman power as a function of coupled pump power (**a**) at different repetition frequencies of the pump pulse when the fiber length is 20 m and the gas pressure is 16 bar and (**b**) at different fiber lengths when the repetition frequency is 1.5 MHz and the gas pressure is 16 bar [29].

4.4. Theoretical Model

Because there are many adjustable experimental parameters in 1.7 μm FGRLs, it is necessary to establish a corresponding theoretical model to guide the actual experiment to obtain a higher output Raman power. In fact, the 1.7 μm fiber hydrogen Raman laser has the following characteristics: First, there is only a pure rotational SRS process in the HC-PCFs; second, there is only one rotational Raman line generated by the rotational SRS; and third, there is only one second-order Raman line generated by the cascade Raman conversion. Therefore, the theoretical model is greatly simplified, and a steady-state coupled wave equation considering cascaded Raman conversion and pulse shape is established [57].

$$\begin{cases} \frac{dI_{S2}}{dz} = g_{S2}I_{S2}I_{S1} - \alpha_{S2}I_{S2} \\ \frac{dI_{S1}}{dz} = g_{S1}I_{S1}I_p - \alpha_{S1}I_{S1} - \frac{v_{S1}}{v_{S2}}g_{S2}I_{S2}I_{S1} \\ \frac{dI_P}{dz} = -\frac{v_P}{v_{S1}}g_{S1}I_{S1}I_P - \alpha_P I_P \end{cases} \quad (2)$$

where I_x is the intensity, α_x is the fiber loss, v_x is the frequency, and g_x is the steady-state Raman gain coefficient (x means S1 for the first-order Stokes wave, S2 for the second-order

Stokes wave, and P for the pump wave); z is the position of the fiber along the propagation. The boundary conditions of Equation (2) are set as follows [57]:

$$\begin{cases} I_P(z=0) = I_o e^{-\frac{t^2}{2\sigma^2}} \\ I_{S1}(z=0) = \frac{h\upsilon_{S1}\pi\Delta\upsilon_R}{A_{eff}} \end{cases} \tag{3}$$

where I_0 is the initial pump intensity, h is the Planck constant, $\Delta\upsilon_R$ is the Raman linewidth, A_{eff} is the mode field area of the HC-PCFs, σ is the variance of the Gaussian distribution, and t is the pump pulse width.

The above theoretical model is used for simulation, and simulation parameters such as fiber length, pump power, and Raman gain are set according to actual experimental conditions. The simulation and measured results of pulse shapes are shown in Figure 10 [57]. Figure 10a,b present the pulse shapes in the condition of no cascaded Raman conversion, and the simulation results reproduce the measured results very well. The center part of the pump pulse is converted into a Raman pulse, so there is a dip in the middle of the residual pump pulse. When the peak power is extremely high, the first-order Raman light is converted into second-order Raman light, so there is also a dip in the middle of the first-order Raman pulse, as shown in Figure 10c,d. The difference between the simulation and the measured results of the Raman pulse shape is caused by the gain accumulation [61], and the gain accumulation is not considered in the simulation as the pump pulse builds up.

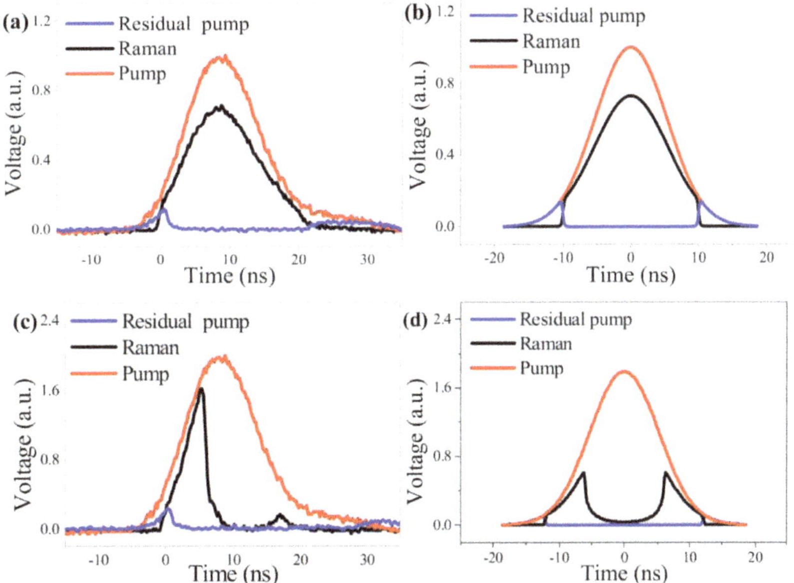

Figure 10. (**a**,**c**) Measured pulse shapes of pump light, first-order Raman light, and residual pump light; (**b**,**d**) corresponding simulation results [57].

Similarly, the simulation evolution curves of the output power with the repetition frequency of the pump pulse are calculated when the fiber length, gas pressure, and average pump power are constant [30], as shown in Figure 11a. The simulation and measured results are plotted using dotted lines and solid-core patterns, respectively. It can be seen that simulation results can find the optical repetition frequency relatively accurately. Furthermore, the simulation evolution curves of the output power with the fiber length are also calculated when the pump power and gas pressure are constant [57], as shown in Figure 11b. It can be seen that the measured results are basically consistent with the simulation results. Thus, the theoretical model is relatively reliable and can have an important guiding role in achieving high-power 1.7 μm FGRLs.

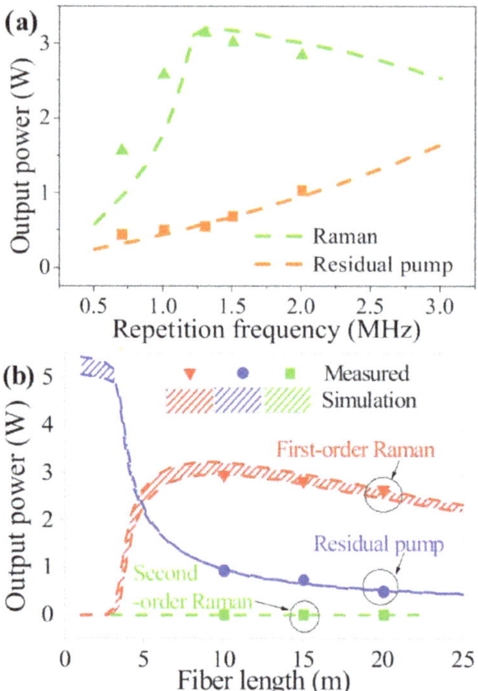

Figure 11. (**a**) The output power as a function of repetition frequency of the pump pulse when the fiber length, gas pressure, and average pump power are constant, and the dotted lines and solid-core patterns represent simulation and measured results, respectively [30]; (**b**) the output power as a function of fiber length when the pump power and gas pressure are constant [57].

5. Summary and Outlook

In summary, we reviewed the studies of 1.7 μm FGRLs based on HC-PCFs. We briefly described the principle and characteristics of HC-PCFs and the gas SRS process and systematically characterized 1.7 μm FGRLs in aspects of output spectral coverage, power-limiting factors, and a theoretical model. There are two important facts to improve the output Raman power for 1.7 μm FGRLs when the fiber length and pump power are constant. One is to keep the gas pressure at a high level to ensure Raman gain saturation in the HC-PCFs. The other is to adjust the parameters of the pump pulse so the peak power is of an appropriate value, that is, the peak power is much larger than the first-order Raman threshold and slightly smaller than the second-order Raman threshold.

We believe that 1.7 μm FGRLs will develop toward higher power and an all-fiber structure. There is no Raman power saturation in the current works, so if more pump power were coupled into the HC-PCFs, a higher output Raman power would be obtained. A useful way is to use a higher-power 1.5 μm pump source and introduce transition fibers to reduce the fusion-splice loss between HC-PCFs and solid-core fibers [62]. To realize an all-fiber structure, it is necessary to solve the problem of HC-PCFs being fusion-spliced with the solid-core fibers in the gas-filled state, which may face the danger of gas leakage and combustion [63]. Furthermore, a 1.7 μm all-fiber gas Raman laser with a Fabry–Perot cavity is also one of the development directions, and the introduction of the Fabry–Perot cavity can further reduce the Raman threshold [64]. However, because H_2 molecules and D_2 molecules are extremely small, they can penetrate the silica glass and overflow from the HC-PCFs, which could affect the long-term reliability of the 1.7 μm FGRL. This problem also needs to be further studied and solved.

Author Contributions: Conceptualization, Z.W.; software, H.L.; investigation, J.L. and H.L.; writing—original draft preparation, J.L.; and writing—review and editing, H.L. and Z.W. All authors have read and agreed to the published version of the manuscript.

Funding: This research was funded by the Outstanding Youth Science Fund Project of the Hunan Province Natural Science Foundation (2019JJ20023), the National Natural Science Foundation of China (NSFC) (11974427, 12004431), and the Fund of State Key Laboratory of Pulsed Power Laser Technology (SKL 2020 ZR05).

Conflicts of Interest: The authors declare no conflict of interest.

References

1. Burns, M.D.; Shardlow, P.C.; Barua, P.; Jefferson-Brain, T.L.; Sahu, J.K.; Clarkson, W.A. 47 W continuous-wave 1726 nm thulium fiber laser core-pumped by an erbium fiber laser. *Opt. Lett.* **2019**, *44*, 5230–5233. [CrossRef] [PubMed]
2. Zhang, Y.; Song, J.; Ye, J.; Xu, J.; Yao, T.; Zhou, P. Tunable random Raman fiber laser at 1.7 μm region with high spectral purity. *Opt. Express* **2019**, *27*, 28800–28807. [CrossRef]
3. Cheng, H.; Tong, S.; Deng, X.; Liu, H.; Du, Y.; He, C. Deep-brain 2-photon fluorescence microscopy in vivo excited at the 1700 nm window. *Opt. Lett.* **2019**, *44*, 4432–4435. [CrossRef] [PubMed]
4. Horton, N.G.; Xu, C. Dispersion compensation in three-photon fluorescence microscopy at 1700 nm. *Biomed. Opt. Express* **2015**, *6*, 1392–1397. [CrossRef] [PubMed]
5. Kawagoe, H.; Ishida, S.; Aramaki, M.; Sakakibara, Y.; Omoda, E.; Kataura, H. Development of a high-power supercontinuum source at the 1.7 μm wavelength region for highly penetrative ultrahigh-resolution optical coherence tomography. *Biomed. Opt. Express* **2014**, *5*, 932–943. [CrossRef]
6. Li, C.; Shi, J.; Gong, X.; Kong, C.; Luo, Z.; Song, L. 1.7 μm wavelength tunable gain-switched fiber laser and its application to spectroscopic photoacoustic imaging. *Opt. Lett.* **2018**, *43*, 5849–5852. [CrossRef]
7. Li, C.; Shi, J.; Wang, X.; Wang, B.; Gong, X.; Song, L. High-energy all-fiber gain-switched thulium-doped fiber laser for volumetric photoacoustic imaging of lipids. *Photonics Res.* **2020**, *8*, 160–164. [CrossRef]
8. Li, B.; Zheng, C.; Liu, H.; He, Q.; Ye, W.; Zhang, Y.; Pan, J.; Wang, Y. Development and measurement of a near-infrared CH_4 detection system using 1.654 μm wavelength-modulated diode laser and open reflective gas sensing probe. *Sens. Actuators B Chem.* **2016**, *225*, 188–198. [CrossRef]
9. Yin, T.; Qi, Z.; Chen, F.; Song, Y.; He, S. High peak-power and narrow-linewidth all-fiber Raman nanosecond laser in 1.65 μm waveband. *Opt. Express* **2020**, *28*, 7175–7181. [CrossRef]
10. Li, C.; Kong, C.; Wong, K.K.Y. High Energy Noise-Like Pulse Generation from a Mode-Locked Thulium-Doped Fiber Laser at 1.7 μm. *IEEE Photonics J.* **2019**, *11*, 1–6. [CrossRef]
11. Li, C.; Wei, X.; Kong, C.; Tan, S.; Chen, N.; Kang, J. Fiber chirped pulse amplification of a short wavelength mode-locked thulium-doped fiber laser. *APL Photonics* **2017**, *2*, 121302. [CrossRef]
12. Du, T.; Ruan, Q.; Yang, R.; Li, W.; Wang, K.; Luo, Z. 1.7-μm Tm/Ho-Codoped All-Fiber Pulsed Laser Based on Intermode-Beating Modulation Technique. *J. Lightwave Technol.* **2018**, *36*, 4894–4899. [CrossRef]
13. Noronen, T.; Okhotnikov, O.; Gumenyuk, R. Electronically tunable thulium-holmium mode-locked fiber laser for the 1700–1800 nm wavelength band. *Opt. Express* **2016**, *24*, 14703–14708. [CrossRef]
14. Noronen, T.; Firstov, S.; Dianov, E.; Okhotnikov, O.G. 1700 nm dispersion managed mode-locked bismuth fiber laser. *Sci. Rep.* **2016**, *6*, 24876. [CrossRef] [PubMed]
15. Khegai, A.; Melkumov, M.; Riumkin, K.; Khopin, V.; Firstov, S.; Dianov, E. NALM-based bismuth-doped fiber laser at 1.7 μm. *Opt. Lett.* **2018**, *43*, 1127–1130. [CrossRef]
16. Thipparapu, N.K.; Wang, Y.; Wang, S.; Umnikov, A.A.; Barua, P.; Sahu, J.K. Bi-doped fiber amplifiers and lasers [Invited]. *Opt. Mater. Express* **2019**, *9*, 2446–2465. [CrossRef]
17. Fang, X.; Wang, Z.Q.; Zhan, L. Efficient generation of all-fiber femtosecond pulses at 1.7 μm via soliton self-frequency shift. *Opt. Eng.* **2017**, *56*, 046107. [CrossRef]
18. Nguyen, T.N.; Kieu, K.; Churin, D.; Ota, T.; Miyawaki, M.; Peyghambarian, N. High Power Soliton Self-Frequency Shift with Improved Flatness Ranging From 1.6 to 1.78 μm. *IEEE Photonics Technol. Lett.* **2013**, *25*, 1893–1896. [CrossRef]
19. Wang, K.; Xu, C. Tunable high-energy soliton pulse generation from a large-mode-area fiber and its application to third harmonic generation microscopy. *Appl. Phys. Lett.* **2011**, *99*, 071112. [CrossRef]
20. Zach, A.; Mohseni, M.; Polzer, C.; Nicholson, J.W.; Hellerer, T. All-fiber widely tunable ultrafast laser source for multimodal imaging in nonlinear microscopy. *Opt. Lett.* **2019**, *44*, 5218–5221. [CrossRef]

21. Becheker, R.; Tang, M.; Hanzard, P.H.; Tyazhev, A.; Mussot, A.; Kudlinski, A. High-energy dissipative soliton-driven fiber optical parametric oscillator emitting at 1.7 μm. *Laser Phys. Lett.* **2018**, *15*, 115103. [CrossRef]
22. Qin, Y.; Batjargal, O.; Cromey, B.; Kieu, K. All-fiber high-power 1700 nm femtosecond laser based on optical parametric chirped-pulse amplification. *Opt. Express* **2020**, *28*, 2317–2325. [CrossRef] [PubMed]
23. Tang, M.; Becheker, R.; Hanzard, P.H.; Tyazhev, A.; Oudar, J.L.; Mussot, A. Low Noise High-Energy Dissipative Soliton Erbium Fiber Laser for Fiber Optical Parametric Oscillator Pumping. *Appl. Sci.* **2018**, *8*, 2161. [CrossRef]
24. Zeng, J.; Akosman, A.E.; Sander, M.Y. Supercontinuum Generation from a Thulium Ultrafast Fiber Laser in a High NA Silica Fiber. *IEEE Photonics Technol. Lett.* **2019**, *31*, 1787–1790. [CrossRef]
25. Chung, H.Y.; Liu, W.; Cao, Q.; Kartner, F.X.; Chang, G. Er-fiber laser enabled, energy scalable femtosecond source tunable from 1.3 to 1.7 μm. *Opt. Express* **2017**, *25*, 15760–15771. [CrossRef] [PubMed]
26. Grimes, A.; Hariharan, A.; Sun, Y.; Ovtar, S.; Kristensen, P.; Westergaard, P.G.; Rako, S.; Baumgarten, C.; Stoneman, R.C.; Nicholson, J.W.; et al. Hundred-watt CW and Joule level pulsed output from Raman fiber laser in 1.7-μm band. In Proceedings of the SPIE 11260, Fiber Lasers XVII: Technology and Systems, San Francisco, CA, USA, 21 February 2020.
27. Cui, Y.; Huang, W.; Li, Z.; Zhou, Z.; Wang, Z. High-efficiency laser wavelength conversion in deuterium-filled hollow-core photonic crystal fiber by rotational stimulated Raman scattering. *Opt. Express* **2019**, *27*, 30396–30404. [CrossRef]
28. Huang, W.; Li, Z.; Cui, Y.; Zhou, Z.; Wang, Z. Efficient, watt-level, tunable 1.7 μm fiber Raman laser in H_2-filled hollow-core fibers. *Opt. Lett.* **2020**, *45*, 475–478. [CrossRef]
29. Li, H.; Huang, W.; Cui, Y.; Zhou, Z.; Wang, Z. 3W tunable 1.65 μm fiber gas Raman laser in D_2-filled hollow-core photonic crystal fibers. *Opt. Laser Technol.* **2020**, *132*, 106474. [CrossRef]
30. Li, H.; Pei, W.; Huang, W.; Cui, Y.; Wang, M.; Wang, Z. Highly efficient nanosecond 1.7 μm fiber gas Raman laser by H_2-filled hollow-core photonic crystal fibers. *Crystals* **2021**, *11*, 32. [CrossRef]
31. Cregan, R.F.; Mangan, B.J.; Knight, J.C.; Birks, T.A.; Russell, P.S.J.; Roberts, P.J.; Allan, D.C. Single-Mode Photonic Band Gap Guidance of Light in Air. *Science* **1999**, *285*, 1537–1539. [CrossRef]
32. Mangan, B.J.; Farr, L.; Langford, A.; Roberts, P.J.; Williams, D.P.; Couny, F.; Lawman, M.; Mason, M.; Coupland, S.; Flea, R. Low loss (1.7 dB/km) hollow core photonic bandgap fiber. In Proceedings of the Optical Fiber Communication Conference, Los Angeles, CA, USA, 22 February 2004.
33. Roberts, P.J.; Couny, F.; Sabert, H.; Mangan, B.J.; Russell, P.S.J. Ultimate low loss of hollow-core photonic crystal fibres. *Opt. Express* **2005**, *13*, 236–244. [CrossRef] [PubMed]
34. Zhang, X.; Gao, S.; Wang, Y.; Ding, W.; Wang, X.; Wang, P. 7-cell hollow-core photonic bandgap fiber with broad spectral bandwidth and low loss. *Opt. Express* **2019**, *27*, 11608–11616. [CrossRef] [PubMed]
35. Smith, C.M.; Venkataraman, N.; Gallagher, M.T.; Muller, D.; West, J.A.; Borrelli, N.F.; Allan, D.C.; Koch, K.W. Low-loss hollow-core silica/air photonic bandgap fibre. *Nature* **2003**, *424*, 657–659. [CrossRef] [PubMed]
36. Benabid, F.; Knight, J.C.; Antonopoulos, G.; Russell, P.S.J. Stimulated Raman scattering in hydrogen-filled hollow-core photonic crystal fiber. *Science* **2002**, *298*, 399–402. [CrossRef]
37. Couny, F.; Benabid, F.; Light, P.S. Large-pitch kagome-structured hollow-core photonic crystal fiber. *Opt. Lett.* **2006**, *31*, 3574–3576. [CrossRef]
38. Couny, F.; Benabid, F.; Roberts, P.J.; Light, P.S.; Raymer, M.G. Generation and photonic guidance of multi-octave optical-frequency combs. *Science* **2007**, *318*, 1118–1121. [CrossRef]
39. Pearce, G.J.; Wiederhecker, G.S.; Poulton, C.G.; Burger, S. Models for guidance in kagome-structured hollow-core photonic crystal fibres. *Opt. Express* **2007**, *15*, 12680–12685. [CrossRef]
40. Debord, B.; Amsanpally, A.; Chafer, M.; Baz, A.; Maurel, M.; Blondy, J.M.; Hugonnot, E.; Scol, F.; Vincetti, L.; Gérôme, F.; et al. Ultralow transmission loss in inhibited-coupling guiding hollow fibers. *Optica* **2017**, *4*, 209–217. [CrossRef]
41. Litchinitser, N.M.; Abeeluck, A.K.; Headley, C. Antiresonant reflecting photonic crystal optical waveguides. *Opt. Lett.* **2002**, *27*, 1592–1594. [CrossRef]
42. Wang, Y.Y.; Wheeler, N.V.; Couny, F. Low loss broadband transmission in hypocycloid-core Kagome hollow-core photonic crystal fiber. *Opt. Lett.* **2011**, *36*, 669–671. [CrossRef]
43. Gérôme, F.; Jamier, R.; Auguste, J.L. Simplified hollow-core photonic crystal fiber. *Opt. Lett.* **2010**, *35*, 1157–1159. [CrossRef] [PubMed]
44. Pryamikov, A.D.; Biriukov, A.S.; Kosolapov, A.F.; Plotnichenko, V.G.; Semjonov, S.L.; Dianov, E.M. Demonstration of a waveguide regime for a silica hollow-core micro structured optical fiber with a negative curvature of the core boundary in the spectral region >3.5 μm. *Opt. Express* **2011**, *19*, 1441–1448. [CrossRef] [PubMed]
45. Yu, F.; Wadsworth, W.J.; Knight, J.C. Low loss silica hollow core fibers for 3–4 μm spectral region. *Opt. Express* **2012**, *20*, 11153–11158. [CrossRef]
46. Kolyadin, A.N.; Kosolapov, A.F.; Pryamikov, A.D.; Biriukov, A.S.; Plotnichenko, V.G.; Dianov, E.M. Light transmission in negative curvature hollow core fiber in extremely high material loss region. *Opt. Express* **2013**, *21*, 9514–9519. [CrossRef] [PubMed]
47. Gao, S.F.; Wang, Y.Y.; Liu, X.L. Bending loss characterization in nodeless hollow-core anti-resonant fiber. *Opt. Express* **2016**, *24*, 14801–14811. [CrossRef]

48. Bradley, T.D.; Jasion, G.T.; Hayes, J.R. Antiresonant Hollow Core Fibre with 0.65 dB/km Attenuation across the C and L Telecommunication bands. In Proceedings of the 45th European Conference on Optical Communication, Dublin, Ireland, 26 September 2019.
49. Jasion, G.T.; Bradley, T.D.; Harrington, K.; Sakr, H.; Chen, Y.; Fokoua, E.N.; Davidson, I.A.; Taranta, A.; Hayes, J.R.; Richardson, D.J. Hollow Core NANF with 0.28 dB/km Attenuation in the C and L Bands. In Proceedings of the 2020 Optical Fiber Communications Conference and Exhibition, San Diego, CA, USA, 8 March 2020.
50. Vincetti, L.; Setti, V. Extra loss due to Fano resonances in inhibited coupling fibers based on a lattice of tubes. *Opt. Express* **2012**, *20*, 14350–14361. [CrossRef]
51. Jaworski, P.; Yu, F.; Maier, R.R.J.; Wadsworth, W.J.; Knight, J.C.; Shephard, J.D.; Hand, D.P. Picosecond and nanosecond pulse delivery through a hollow-core Negative Curvature Fiber for micro-machining applications. *Opt. Express* **2013**, *21*, 22742–22753. [CrossRef]
52. Bufetov, I.A.; Kosolapov, A.F.; Pryamikov, A.D.; Gladyshev, A.V.; Kolyadin, A.N.; Krylov, A.A.; Yatsenko, Y.P.; Biriukov, A.S. Revolver Hollow Core Optical Fibers. *Fibers* **2018**, *6*, 39. [CrossRef]
53. Komanec, M.; Dousek, D.; Suslov, D.; Zvanove, S. Hollow-Core Optical Fibers. *Radioengineering* **2020**, *29*, 417–430. [CrossRef]
54. Minck, R.W.; Terhune, R.W.; Rado, W.G. Laser stimulated Raman effect and resonant four-photon interactions in gases H_2, D_2, and CH_4. *Appl. Phys. Lett.* **1963**, *3*, 181–183. [CrossRef]
55. Brink, D.J.; Proch, D. Efficient tunable ultraviolet source based on stimulated Raman scattering of an excimer-pumped dye laser. *Opt. Lett.* **1982**, *7*, 494–496. [CrossRef] [PubMed]
56. Loree, T.R.; Cantrell, C.D.; Barker, D.L. Stimulated Raman emission at 9.2 μm from hydrogen gas. *Opt. Commun.* **1976**, *17*, 160–162. [CrossRef]
57. Li, H.; Huang, W.; Cui, Y.; Zhou, Z.; Wang, Z. Pure rotational stimulated Raman scattering in H_2-filled hollow-core photonic crystal fibers. *Opt. Express* **2020**, *28*, 23881–23897. [CrossRef]
58. Teal, G.K.; MacWood, G.E. The Raman Spectra of the Isotopic Molecules H_2, HD, and D_2. *J. Chem. Phys.* **1935**, *3*, 760–764. [CrossRef]
59. Aghaie, K.Z.; Digonnet, M.J.F.; Fan, S. Optimization of the splice loss between photonic-bandgap fibers and conventional single-mode fibers. *Opt. Lett.* **2010**, *35*, 1938–1940. [CrossRef]
60. Bischel, W.K.; Dyer, M.J. Wavelength dependence of the absolute Raman gain coefficient for the Q(1) transition in H_2. *J. Opt. Soc. Am. B* **1986**, *3*, 677–682. [CrossRef]
61. Benabid, F.; Antonopoulos, G.; Knight, J.C.; Russell, P.S.J. Stokes amplification regimes in quasi-cw pumped hydrogen-filled hollow-core photonic crystal fiber. *Phys. Rev. Lett.* **2005**, *95*, 213903. [CrossRef]
62. Gao, S.; Wang, Y.; Tian, C.; Wang, P. Splice Loss Optimization of a Photonic Bandgap Fiber via a High V-Number Fiber. *IEEE Photonics Technol. Lett.* **2014**, *26*, 2134–2137. [CrossRef]
63. Benabid, F.; Couny, F.; Knight, J.C.; Birks, T.A.; Russell, P.S. Compact, stable and efficient all-fibre gas cells using hollow-core photonic crystal fibres. *Nature* **2005**, *434*, 488–491. [CrossRef]
64. Couny, F.; Benabid, F.; Light, P.S. Subwatt threshold cw Raman fiber-gas laser based on H_2-filled hollow-core photonic crystal fiber. *Phys. Rev. Lett.* **2007**, *99*, 143903. [CrossRef]

Article

Hydrogen Molecules Rotational Stimulated Raman Scattering in All-Fiber Cavity Based on Hollow-Core Photonic Crystal Fibers

Wenxi Pei [1,2], Hao Li [1,2], Wei Huang [1,3], Meng Wang [1,2,3] and Zefeng Wang [1,2,3,*]

1 College of Advanced Interdisciplinary Studies, National University of Defense Technology, Changsha 410073, China; peiwenxi@nudt.edu.cn (W.P.); lihao18c@nudt.edu.cn (H.L.); cuiyulong@nudt.edu.cn (W.H.); wangmeng@nudt.edu.cn (M.W.)
2 Hunan Provincial Key Laboratory of High Energy Laser Technology, Changsha 410073, China
3 State Key Laboratory of Pulsed Power Laser Technology, Changsha 410073, China
* Correspondence: zefengwang@nudt.edu.cn

Abstract: Here, we report the rotational stimulated Raman scattering (SRS) of hydrogen molecules in an all-fiber cavity based on hollow-core photonic crystal fibers (HC-PCFs). The gas cavity consists of a 49 m long HC-PCF filled with 18 bar high-pressure hydrogen and two sections of fusion spliced solid-core fibers on both ends. When pumped by a homemade 1064 nm pulsed fiber amplifier, only rotational SRS occurs in the gas cavity due to the transmission spectral characteristics of the used HC-PCF, and 1135 nm Stokes wave is obtained (Raman frequency shift of 587 cm^{-1}). By changing the pulse width and repetition frequency of the pump source, the output characteristics are explored. In addition, a theoretical model is established for comparison with the experimental results. This work is helpful for the application of gas Raman laser based on the HC-PCFs.

Keywords: photonic crystal fiber; hollow-core fiber; fiber gas laser; stimulated Raman scattering

Citation: Pei, W.; Li, H.; Huang, W.; Wang, M.; Wang, Z. Hydrogen Molecules Rotational Stimulated Raman Scattering in All-Fiber Cavity Based on Hollow-Core Photonic Crystal Fibers. *Crystals* **2021**, *11*, 711. https://doi.org/10.3390/cryst11060711

Academic Editors: David Novoa and Nicolas Y. Joly

Received: 22 May 2021
Accepted: 18 June 2021
Published: 21 June 2021

1. Introduction

Since the first demonstration in 1963 [1], stimulated Raman scattering (SRS) in gases has been considered as an effective way of generating lasers at unobtainable wavelengths. Due to the large Raman shift and high Raman gain coefficient, light sources based on the SRS in gases can obtain tunable, narrow linewidth lasers, especially for the lasers in the ultraviolet and infrared spectral range [2,3]. However, the SRS in the gas cell has a short interaction length, which requires very high pump peak power, usually up to megawatt level. Besides, there are always more than one Raman line generated and the Raman conversion efficiency is quite low for the lasers at the desired wavelength. The appearance of hollow-core fibers (HCFs) provides an effective solution to these problems and gives birth to a new kind of fiber lasers, namely, fibers gas Raman lasers (FGRLs). The gas cell based on HCFs can greatly enhance the interaction between the pump laser and the gas molecules. The interaction length can be also greatly extended by increasing the length of the HCFs. What is more, the final gain spectrum of various orders of Stokes waves can be controlled by wavelength-dependent fiber attenuation in the HCFs, which can provide much higher conversion efficiency of the desired wavelength. Gaseous hydrogen is the most commonly used Raman gain medium in gas cells. The first demonstration of the SRS of hydrogen molecules in HCFs was in 2002 [4], then a lot of work has been done based on various HCFs [4–15]. However, in almost all reported FGRLs, the pump laser is coupled into the HCFs through free space, which is quite inconvenient and limits the applications of the FGRLs. Thus, in 2004 [13], Benabid et al. first reported an all-fiber gas cell spliced by an expensive special fiber fusion splicer (Vytran FFS-2000-PM, America), which is an effective but not common way. In 2007 [6], Couny et al. reported an all-fiber gas cavity, but the pump laser was coupled into the cavity through free space. Therefore, they did not realize the experimental system with all-fiber structure. In our previous work [11,15], we reported the 1.7 μm quasi all-fiber FGRL, and the optical isolators were introduced to avoid Fresnel reflection damage to the pump source. The fiber laser was used as the pump source and the pigtail was spliced to the input end of the HC-PCF, which greatly improves the stability of the system. However, the output end of the HC-PCF is still sealed in the gas chamber, making the experimental system still bulky. Chen et al. reported a 1135 nm FGRL with all-fiber structure in 2013 [14], but the power is less than 1 mW and the conversion efficiency is only 2.69%, meaning poor performance, and it is quite difficult to explore the applications.

In this paper, we report the hydrogen SRS in an all-fiber cavity based on HC-PCFs. The gas cavity is fabricated by splicing the HC-PCFs with two sections of solid-core fibers at each end using arc fusion technique [16–18]. When pumped with a pulsed 1064 nm fiber laser amplifier, a 1135 nm Stokes wave is generated by pure rotational SRS of hydrogen molecules due to the transmission loss spectrum of the HC-PCF. The output characteristics are explored with different pump pulse widths and repetition frequencies. When the width is 4 ns and the repetition frequency is 20 MHz, the maximum Stokes power of 289 mW is achieved with a power conversion efficiency of 28.5%, which can be greatly enhanced by improving the splicing loss and optimizing the HCF's length in the future. A theoretical model is established, and the experimental result is in good agreement with the simulation result.

2. Experimental Setup

Figure 1 shows the experimental setup of pulsed 1 μm all-fiber Raman laser. The pump source consists of a seed which is a 1064 nm optical pulse machine (OPM; a laser diode with adjustable pulse generator, pulse driver, and efficient TEC controller) with multi-longitudinal modes and a homemade double-cladding ytterbium-doped fiber amplifier (YDFA). The pulse width of the seed can be adjusted form 1 ns to 6 ns and the repetition frequency tuning range is from 4 kHz to 25 MHz. The maximum average output power of 52.42 mW can be achieved when the seed is at the pulse width of 4 ns and the repetition frequency is 20 MHz. Figure 2a presents the measured fine spectrum with an OSA resolution of 0.02 nm in this power, and it can be seen that there are four longitudinal modes. In YDFA, there are 5 m ytterbium-doped fibers (YDF; Nufern double clad optical fiber, USA) with the absorption coefficient of 4.95 dB/m at 976 nm and the core diameter is 10 μm. The YDF is pumped by a 976 nm laser diode (LD) with the maximum output power of 3.28 W, which is coupled into the YDF by the fiber combiner. YDFA is fused with an optical isolator (the insertion loss is 0.46 dB, the isolation is 38.5 dB) to avoid the disturbance caused by the backward laser. The maximum output power of 1.51 W is achieved with the 3.28 W pump power at 976 nm when the pulse width of seed is 4 ns, and the repetition frequency is 20 MHz. The measured fine spectrum of the YDFA in Figure 2b keeps the characteristics well, and the corresponding wavelengths are 1063.94 nm, 1064.38 nm, 1064.83 nm, and 1065.27 nm, respectively. The output fiber of the isolator is fused with a fiber coupler (the measured coupling ratio is 99.83:0.17) to monitor the output power of the pump source in real time. The main output fiber of the coupler is fused with the all-fiber gas cavity. A convex-plane lens is placed at the output end of the gas cavity to collimate the laser. Two silver mirrors can help monitor the output spectrum. Besides, the residual pump laser and Stokes laser are separated by adjusting the flippable long-pass filters, which is helpful to measure the power.

For the gas cavity, a 49 m long HC-PCF (NKT Photonics HC-1060-02, Denmark) filled with 18 bar hydrogen is spliced to a solid-core fiber (Corning HI1060, USA) as the pump injection end and the other end of the HC-PCF is spliced to another solid-core fiber (SM-GDF-10/125-15FA) as the output end. In our work, the end as the pump injection is spliced first and the other end of the HC-PCF is sealed into a specially designed gas chamber, which can keep the HC-PCF vacuumed and fill it with hydrogen. After standing for 10 h, the HC-PCF is taken out from the gas chamber and spliced with the solid-core fiber quickly. By calculating the amount of gas leakage in the fusion splicing process [16], the gas pressure in the all-fiber gas cavity is estimated to be 18 bar. Besides, the fusion loss of mode field mismatch can be reduced by using these different solid-core fibers, and the fusion losses of splice1 and splice2 is 1.56 dB and 0.45 dB, respectively. Figure 3a presents the cross-section of the HC-PCF under the electron microscope (the core diameter is 10 μm), and Figure 3b presents the transmission loss and dispersion curves (from the product manual provided by NKT Photonics). It can be seen that the losses at 1064 nm and 1135 nm are ~0.083 dB/m and ~0.012 dB/m, respectively.

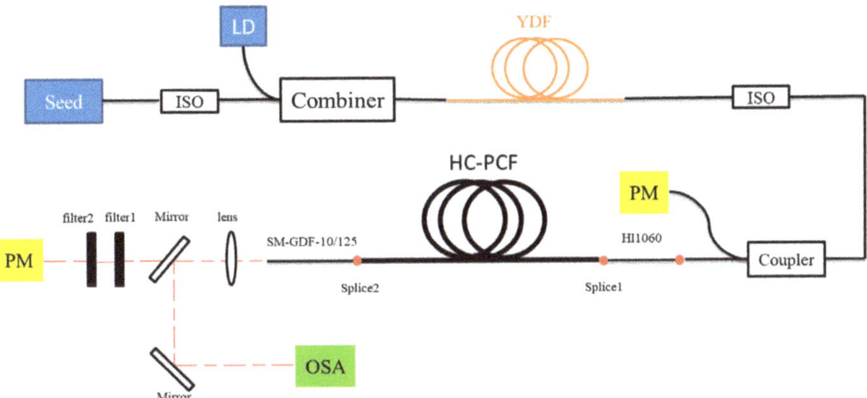

Figure 1. Experimental setup: ISO, optical isolator, LD, laser dioxide, PM, power meter, OSA, optical spectrum analyzer.

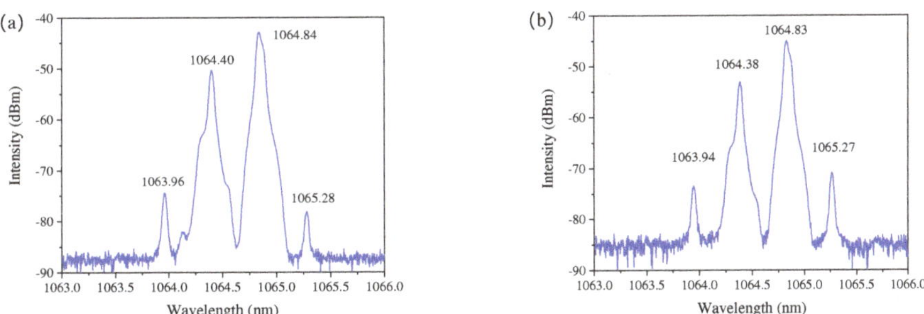

Figure 2. The fine spectrum of (**a**) the seed and (**b**) the fiber amplifier at the maximum output power with an OSA resolution of 0.02 nm.

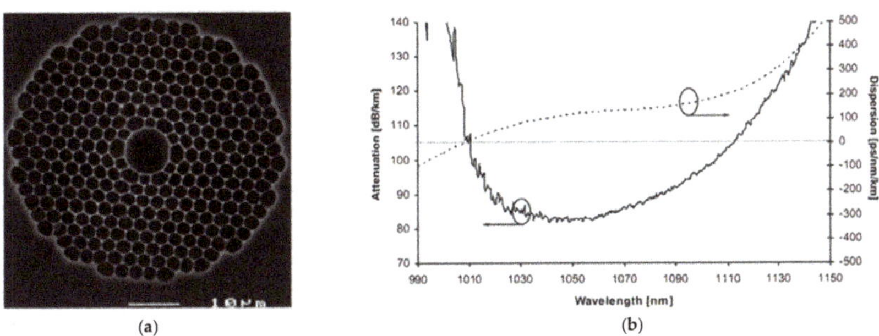

Figure 3. (**a**) The schematic cross section of the HC-PCF and (**b**) the transmission loss and dispersion curves of the HC-PCF, both from the product manual provided by NKT Photonics.

3. Experimental Results and Discussion

The output spectrum is measured when the pump source is at the maximum output power of 1.51 W, as shown in Figure 4a. The width of the pump pulse is 4 ns and the repetition frequency is 20 MHz. It can be seen that besides the pump line at 1064 nm, there are two lines in the spectrum. The line at 1135 nm is the first-order Raman line, and it is converted from 1064 nm pump line by the rotational SRS of hydrogen molecules in the gas cavity (the corresponding Raman frequency shift is 587 cm^{-1}). The line at 1217 nm is the second-order Raman line converted from 1135 nm first-order Raman line with the same Raman frequency shift. Figure 4b–d presents the fine output spectrum of the residual pump laser, the first-order Stokes wave, and the second-order Stokes wave, respectively. The spectrum of the residual pump laser is similar to the output spectrum of the pump source in Figure 2b, but the intensity of some wavelengths has changed. The lines at 1064.38 nm and 1064.83 nm become weaker as they are strong enough to be converted into the first-order Stokes waves; it can be well verified from Figure 4c. There are two lines at 1135.32 nm and 1135.86 nm in the spectrum, which represent two different longitudinal modes. They are generated by the rotational SRS of hydrogen molecules pumped by the lasers at 1064.38 nm and 1064.83 nm, respectively. For the second-order Stokes wave at 1217 nm, there is only one Raman line in the spectrum as shown in Figure 4d. It is converted by the first-order Stokes wave of 1135.86 nm as it is strong enough.

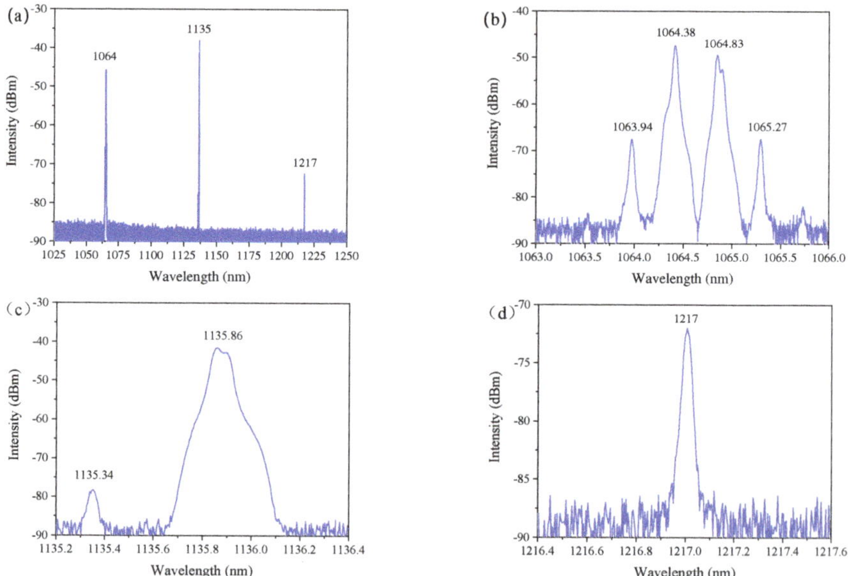

Figure 4. (**a**) The total output spectrum of 1 µm all-fiber FGRL. (**b–d**) The fine spectrum of the residual pump laser, the first-order Stokes laser, and the second-order Stokes laser, respectively.

To explore the impact of the pulse width on the output characteristics of the all-fiber FGRL, we change the pump pulse width from 4 ns to 2 ns and keep the repetition frequency at 20 MHz. Figure 5a shows the output spectrum of the pump source with the maximum output power. It can be seen that there are many longitudinal modes. Figure 5b presents the output spectrum of the FGRL. There are mainly two lines in the output spectrum: a 1064 nm pump line and 1135 nm Raman line. Therefore, only the first-order Stokes wave is generated by the rotational SRS of hydrogen molecules, which is quite different from the case when the pump pulse width is 4 ns. There is a very weak line at 1106 nm that can only be detected by the highly sensitive OSA. It is also the first-order Stokes wave converted by the 1064 nm pump laser, and the corresponding Raman frequency shift is

354 cm^{-1}. The corresponding Raman gain of 1106 nm is weaker than that in 1135 nm [19], so the pump laser is converted into a 1135 nm Stokes wave first, and only a few pump lasers are converted into 1106 nm Stokes waves. Figure 5c,d shows the fine spectrum of the first-order Stokes waves at 1135 nm and 1106 nm, respectively, which indicates that only the pump lasers at 1064.38 nm and 1063.94 nm are converted into the first-order Stokes waves (the Stokes wave at 1106.1 nm is converted by the pump laser at 1064.38 nm). Due to the accuracy of spectrometer, the linewidth of these lasers cannot be measured accurately, and some Stokes waves are too weak to be measured. Compared with the pump source with the pulse width of 4 ns, there are more longitudinal modes in the pump source when the pulse width is 2 ns, which may lead to the dispersion of power and the Raman conversion is not as sufficient as the case with the pulse width of 4 ns when the pump power is the same. The absence of the second-order Stokes wave indicates that the first-order Stokes wave is not strong, which also means the conversion is not sufficient.

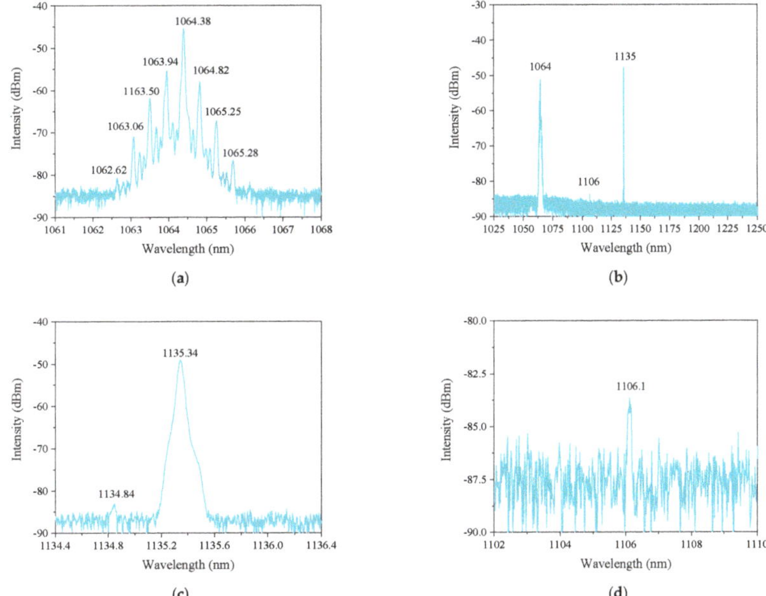

Figure 5. (**a**) The output spectrum of the pump source at the maximum output power when the pulse width of the pump laser is 2 ns and the repetition frequency is 20 MHz. (**b**) The total output spectrum of 1 μm all-fiber FGRL in this case. (**c**,**d**) The fine spectrum of the first-order Stokes waves at 1135 nm and 1106 nm, respectively.

To display more clearly the power components of longitudinal modes, the spectral intensity in Figures 2b and 5a is represented by a linear scale, as shown in Figure 6a,b. It can be seen that although the output spectrum contains many longitudinal modes when the pump pulse width is 2 ns, the power spectrum mainly contains four components with different wavelengths. The pump lines at 1064.38 nm and 1063.94 nm take up the dominant position with the ratio of 77.85% of the total output power. The power spectrum in Figure 6b indicates that when the pulse is 4 ns, the output power has only two main pump lines, and both can be converted into Stokes waves. By calculation, the lines at 1064.83 nm and 1064.38 nm account for 90.03% of the total output power, which is much higher than the case with the pulse width of 2 ns. Therefore, when the pump source is in the same output power and the pulse width is 4 ns, there are more pump lines with higher power participate in Raman conversion. This also explains why the second-order Stokes wave is not generated with the same pump power when the pump pulse is 2 ns.

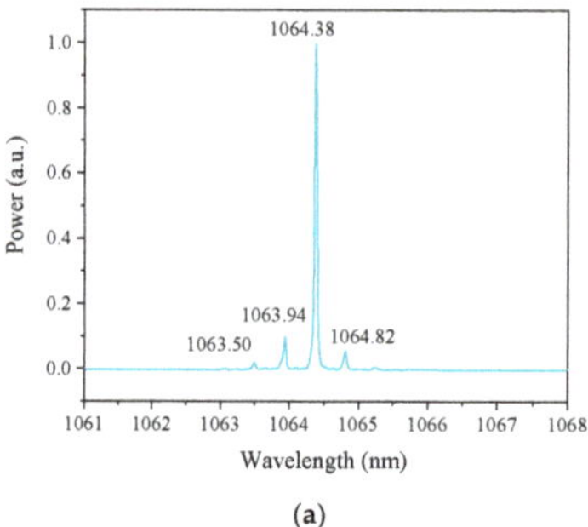

(a)

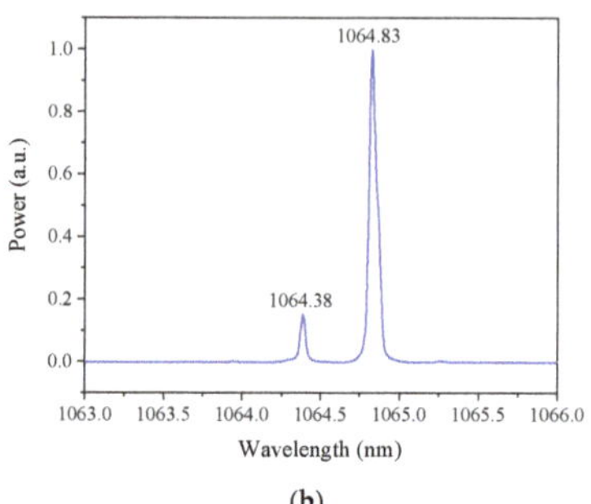

(b)

Figure 6. (**a**,**b**) The power spectrums of the pump source at the maximum power with the pulse width of 2 ns and 4 ns, respectively.

The output power characteristics of the FGRL with different pulse width of pump laser are measured, as shown in Figure 7a–c. The solid and hollow patterns in Figure 7a,b represent the Raman power and residual pump power, respectively. Moreover, the coupled pump power is the actual power that transmitted into the gas cavity after the Splice 1. It can be seen that with the increase in the coupled pump power, the Raman power with the pulse width of 2 ns is detected first. This is because the pump pulse with width 2 ns is narrower, which means the corresponding peak power is higher than the pulse with the width of 4 ns when the pump power is the same and exceed the Raman threshold earlier. Both the Raman powers with different pulse widths increase linearly. The residual pump power with the pulse width of 4 ns drops when the coupled pump power exceeds the Raman threshold, while it keeps increasing when the pulse is 2 ns. It indicates the Raman conversion is not sufficient when the pulse width is 2 ns in the case. The corresponding Raman efficiency (the ratio of coupled pump power to Raman power in the gas cavity) is calculated, as shown in Figure 7c. The maximum Raman efficiency 28.5% is obtained with the maximum coupled pump power and the pulse width is 4 ns. Besides, the Stokes pulse energy and the coupled pulse energy are explored in the Figure 7d. It can be seen that the Stokes energy with the pulse width of 4 ns is higher in general.

Furthermore, we measured the output power characteristics of the FGRL with different repetition frequencies and the same pulse width of 4 ns. Except for the repetition frequencies, the characteristics of the pump pulse like duration and spectral content are the same. Figure 8a,b presents the evolution of Raman power, residual pump power, and Raman efficiency with the increase of coupled pump power, respectively. It can be seen from Figure 8a,b that the curves with different repetition frequencies show similar trends. All the Raman powers increase linearly, while all the residual powers increase at first and then drop when the coupled pump power exceed the Raman threshold. Obviously, the optimal repetition frequency is 20 MHz with the lowest Raman threshold and the highest Raman power among the five repetition frequencies. When the repetition frequency is less than 20 MHz, the peak powers of the pump pulse and Raman pulse will become higher, which is conducive to the conversion of the second-order Raman laser. When it is higher than 20 MHz, less energy of pump pulses that exceed the first-order Raman threshold are converted into Raman pulses. Both situations will reduce the first-order Raman conversion efficiency. Figure 8c depicts the Raman efficiencies with different repetition frequencies. The maximum Raman efficiency is 28.5% in 20 MHz as mentioned earlier, which is more efficient than the previous work [14]. The corresponding Raman power is 289 mW, which is the maximum output Raman power. Compared with the previous work [14], the output power is increased by an order of magnitude. The pulse energy of Stokes pulse and coupled pump pulse is explored as shown in Figure 8d. It can be seen when the repetition frequency is 20 MHz, the Stokes pulse has the highest energy.

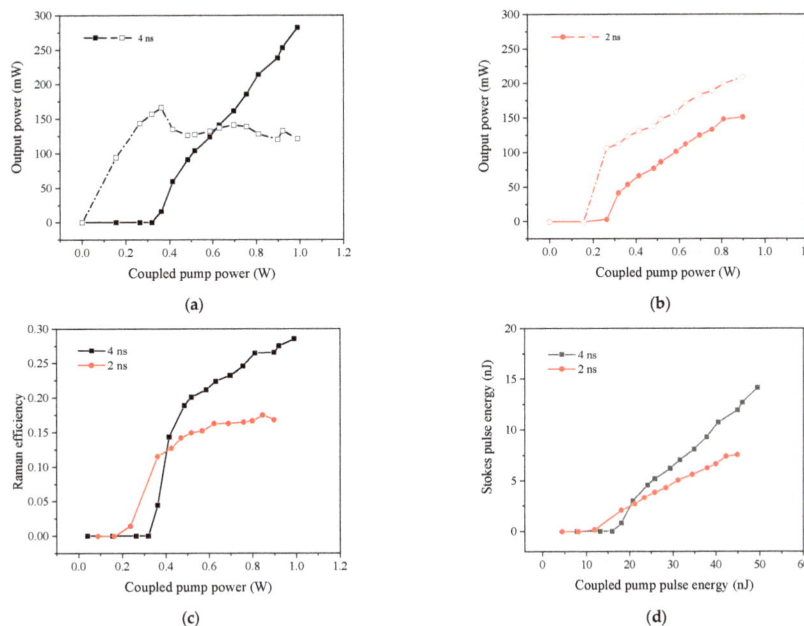

Figure 7. The output power characteristics of the FGRL with the pulse width of (**a**) 4 ns and (**b**) 2 ns. The solid and hollow patterns represent the Raman power and residual pump power, respectively. (**c**) The Raman efficiency of different pulse widths. (**d**) The Stokes pulse energy with coupled pump pulse energy in different repetition frequency.

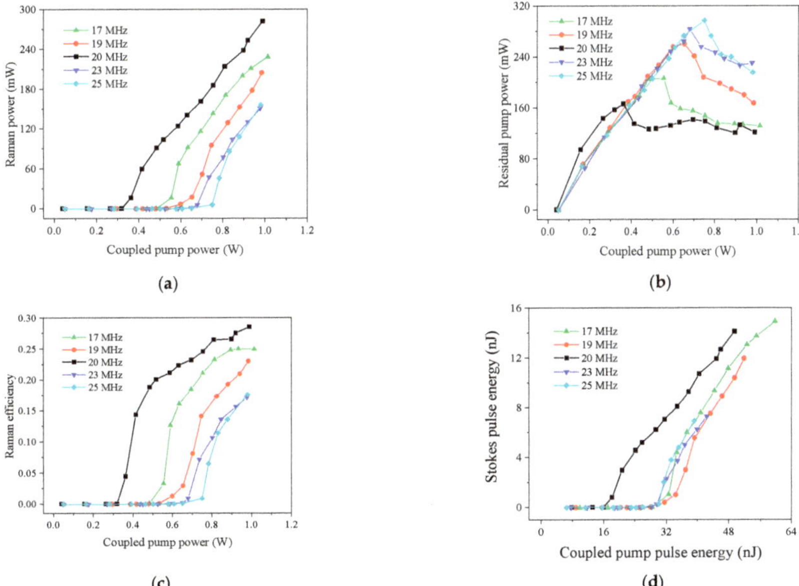

Figure 8. The output power characteristics of the FGRL with different repetition frequencies (pulse width 4 ns). (**a**–**c**) The Raman power, residual pump power and Raman efficiency, respectively. (**d**) The Stokes pulse energy with coupled pump pulse energy in different repetition frequency.

Figure 9a–d presents the shapes and the series of the pump pulses and the Stokes pulses when the pump power is maximum and the repetition is 20 MHz. It can be seen that the frequency of the Stokes pulses is 20 MHz, which is the same as the pump pulses. The pulse widths are measured, as shown in Figure 9a,c. It can be seen that the width of the Stokes pulse is narrower than that of pump pulse. This is not unexpected, because only the center part of the pump pulse that is higher than Raman threshold can be converted into the Stokes pulses.

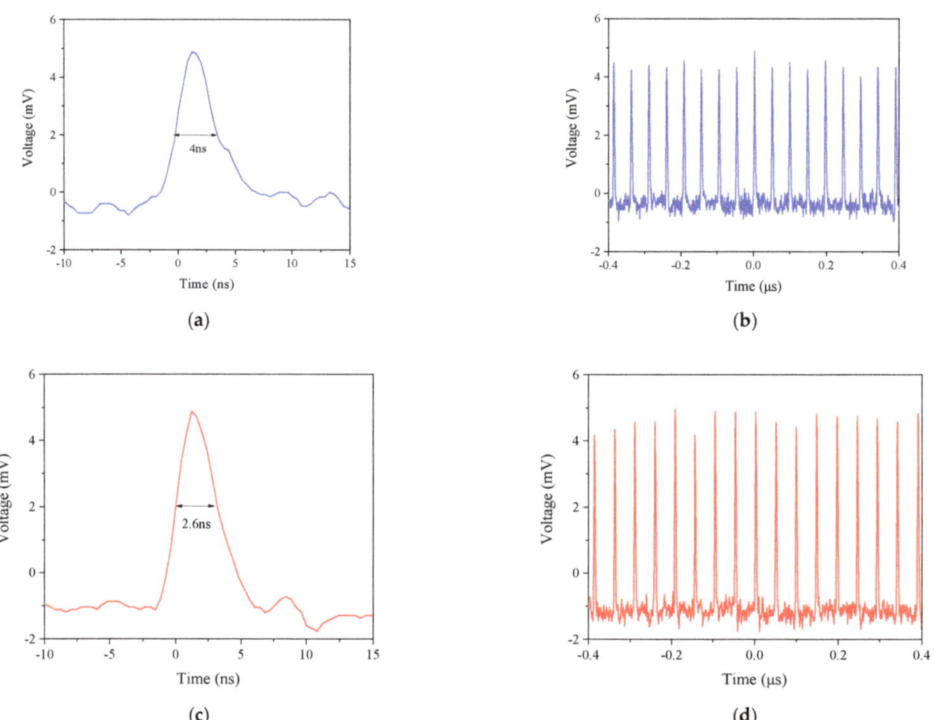

Figure 9. The temporal characteristic of the FGRL when the pump source is at the maximum power and the repetition is 20 MHz. (**a**,**b**) The shapes and the series of the pump pulses. (**c**,**d**) The shapes and the series of the Stokes pulses.

For the single FGRL in our work, the process of the rotational SRS that generates the first-order and the second-order Stokes waves obey the rules as follows [15]:

$$\frac{dP_P}{dz} = -\frac{\lambda_{S1}}{\lambda_P} g_{S1} P_{S1} P_P - \alpha_P P_P \tag{1}$$

$$\frac{dP_{S1}}{dz} = g_{S1} P_{S1} P_P - \alpha_{S1} P_{S1} - \frac{\lambda_{S2}}{\lambda_{S1}} g_{S2} P_{S2} P_{S1} \tag{2}$$

$$\frac{dP_{S2}}{dz} = g_{S2} P_{S2} P_{S1} - \alpha_{S2} P_{S2} \tag{3}$$

where the subscripts of P, $S1$, and $S2$ represent the pump wave, the first-order Stokes wave, and the second-order Stokes wave, respectively; z is the coordinate of the HC-PCF length; λ is the wavelength; g is the steady-state Raman gain coefficient; P is the intensity; and α is the fiber loss. The boundary conditions can be given [11]:

$$P_P(z=0) = \frac{P_0}{A_{eff}} \tag{4}$$

$$P_{s1}(z=0) = \frac{hc\pi\Delta\nu_R}{\lambda_{S1}A_{eff}} \tag{5}$$

where P_0 is the peak power of the coupled pump pulse in the HC-PCFs; A_{eff} is the mode field area of the HC-PCFs; h is the Planck constant; c is the light speed $\Delta\nu_R$; is the Raman linewidth. A simple steady-state theoretical model can be established by the formulas above.

In our work, Figure 10a presents the simulation result of the pump pulse shape and the first-order Stokes pulse shape based on the theoretical model. Here $\Delta\nu_R$ is 3.1 GHz [20]. And g_{S1} is 0.35 cm/GW and g_{S2} is 0.18 cm/GW, which is estimated and adjusted slightly from in [10,11,18]. The mode field diameter of the HC-PCF is 7.5 µm and A_{eff} is 14.06π µm^2. The pump pulse width is 4 ns and the repetition frequency is 20 MHz. It can be seen the Stokes pulse width is ~2.8 ns, which is in good agreement with the experimental results. Figure 10b shows the simulation results of the output power characteristics in the same case. For better comparison, the experimental results are plotted with solid patterns, while the simulation results are plotted with dotted lines. We can find that both the results show the similar trend and with the increase of the coupled pump power, the experimental results are in good agreement with the simulation results at first, then there is a deviation between them. We speculated that the homemade pulsed YDFA may has some continuous wave components which are useless in the Raman conversion and its power increases with the increase of coupled pump power. We hypothesize that by optimizing the performance of the amplifier, we can achieve better agreement between the experimental results and the simulation results.

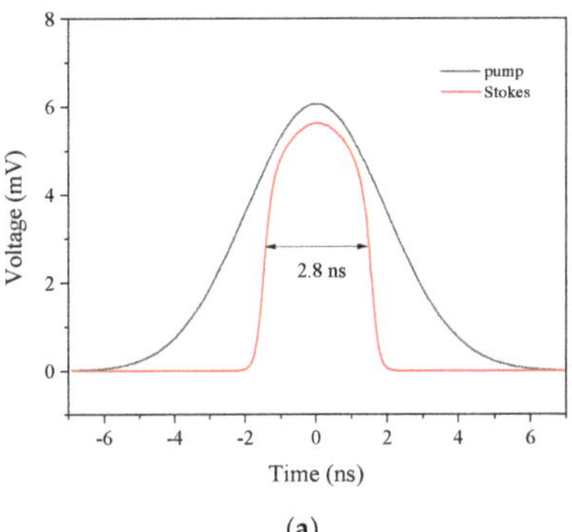

(a)

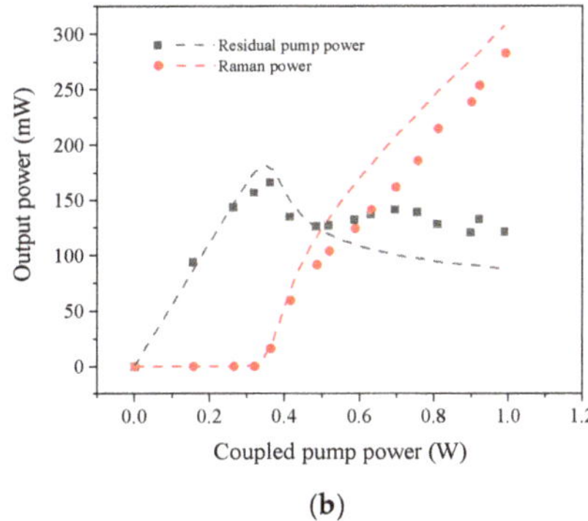

(b)

Figure 10. The simulation results of (**a**) the pump pulse shape and the first-order Stokes pulse shape and (**b**) the output power characteristics.

4. Conclusions

We have demonstrated the rotational SRS of hydrogen molecules in an all-fiber cavity based on HC-PCFs. The all-fiber gas cavity is fabricated by fusion splicing 49 m long HC-PCFs filled with 18 bar hydrogen and two sections of solid-core fibers. When pumped by a homemade 1064 nm YDFA, the pure rotational SRS of hydrogen molecules occurs in the gas cavity. The output characteristics of the FGRL with different repetition frequency and pulse width are explored. The maximum output power with 289 mW of the first-order Stokes wave at 1135 nm is obtained and the corresponding Raman efficiency is 28.5%. A steady-state theoretical model is established, and the experimental results are in good agreement with the simulation results. This work makes a further exploration for the applications of the FGRLs.

Author Contributions: Conceptualization, Z.W. and W.P.; methodology, W.P.; software, W.H. and H.L.; validation, Z.W., H.L. and W.P.; formal analysis, Z.W.; investigation, W.P.; resources, M.W.; data curation, H.L.; writing—original draft preparation, W.P.; writing—review and editing, Z.W.; visualization, W.P. All authors have read and agreed to the published version of the manuscript.

Funding: This work was supported by Outstanding Youth Science Fund Project of Hunan Province Natural Science Foundation (2019JJ20023), National Natural Science Foundation of China (NSFC) (11974427, 12004431) and State Key Laboratory of Pulsed Power Laser Technology (SKL 2020 ZR05).

Institutional Review Board Statement: Not applicable.

Informed Consent Statement: Not applicable.

Data Availability Statement: The data presented in this study are available on request from the corresponding author. The data are not publicly available due to privacy.

Conflicts of Interest: The authors declare no conflict of interest.

References

1. Minck, R.W.; Terhune, R.W.; Rado, W.G. Laser stimulated Raman effect and resonant four-photon interactions in gases H_2, D_2, and CH_4. *Appl. Phys. Lett.* **1963**, *3*, 181–184. [CrossRef]
2. Brink, D.J.; Proch, D. Efficient Tunable Ultraviolet Source Based on Stimulated Raman Scattering of an Excimer-Pumped Dye Laser. *Opt. Lett.* **1982**, *7*, 494–496. [CrossRef] [PubMed]
3. Loree, T.R.; Cantrell, C.D.; Barker, D.L. Stimulated Raman emission at 9.2 μm from hydrogen gas. *Opt. Comm.* **1976**, *17*, 160–162. [CrossRef]
4. Benabid, F. Stimulated Raman Scattering in Hydrogen-Filled Hollow-Core Photonic Crystal Fiber. *Science* **2002**, *298*, 399–402. [CrossRef] [PubMed]
5. Benabid, F.; Bouwmans, G.; Knight, J.C.; Russell, P.S.J.; Couny, F. Ultrahigh Efficiency Laser Wavelength Conversion in a Gas-Filled Hollow Core Photonic Crystal Fiber by Pure Stimulated Rotational Raman Scattering in Molecular Hydrogen. *Phys. Rev. Lett.* **2004**, *93*, 123903. [CrossRef] [PubMed]
6. Couny, F.; Benabid, F.; Light, P.S. Subwatt Threshold Cw Raman Fiber-Gas Laser Based on H_2-Filled Hollow-Core Photonic Crystal Fiber. *Phys. Rev. Lett.* **2007**, *99*, 143903. [CrossRef]
7. Wang, Z.; Yu, F.; Wadsworth, W.J.; Knight, J.C. Efficient 1.9 μm Emission in H_2 -Filled Hollow Core Fiber by Pure Stimulated Vibrational Raman Scattering. *Laser Phys. Lett.* **2014**, *11*, 105807. [CrossRef]
8. Gladyshev, A.V.; Bufetov, I.A.; Dianov, E.M.; Kosolapov, A.F.; Khudyakov, M.M.; Yatsenko, Y.P.; Kolyadin, A.N.; Krylov, A.A.; Pryamikov, A.D.; Biriukov, A.S.; et al. 2.9, 3.3, and 3.5 μm Raman Lasers Based on Revolver Hollow-Core Silica Fiber Filled by $^1H_2/D_2$ Gas Mixture. *IEEE J. Select. Top. Quantum Electron.* **2018**, *24*, 0903008. [CrossRef]
9. Astapovich, M.S.; Gladyshev, A.V.; Khudyakov, M.M.; Kosolapov, A.F.; Likhachev, M.E.; Bufetov, I.A. Watt-Level Nanosecond 4.42 μm Raman Laser Based on Silica Fiber. *IEEE Photon. Technol. Lett.* **2019**, *31*, 78–81. [CrossRef]
10. Huang, W.; Li, Z.; Cui, Y.; Zhou, Z.; Wang, Z. Efficient, Watt-Level, Tunable 1.7 μm Fiber Raman Laser in H_2-Filled Hollow-Core Fibers. *Opt. Lett.* **2020**, *45*, 475–478. [CrossRef]
11. Li, H.; Huang, W.; Cui, Y.; Zhou, Z.; Wang, Z. Pure Rotational Stimulated Raman Scattering in H_2-Filled Hollow-Core Photonic Crystal Fibers. *Opt. Express* **2020**, *28*, 23881–23897. [CrossRef] [PubMed]
12. Thapa, R.; Knabe, K.; Corwin, K.L.; Washburn, B.R. Arc Fusion Splicing of Hollow-Core Photonic Bandgap Fibers for Gas-Filled Fiber Cells. *Opt. Express* **2006**, *14*, 9576–9583. [CrossRef] [PubMed]
13. Benabid, F.; Couny, F.; Knight, J.C.; Birks, T.A.; Russell, P.S.J. Compact, Stable and Efficient All-Fibre Gas Cells Using Hollow-Core Photonic Crystal Fibres. *Nature* **2005**, *434*, 488–491. [CrossRef]
14. Chen, X.D.; Sun, Q.; Li, H.; Zhao, M.H. Compact All-Fiber Gas Raman Light Source Based on Hydrogen-Filled Hollow-Core Photonic Crystal Fiber Pumped with Single-Mode Q-Switched Fiber Laser. *Opt. Fiber Technol.* **2013**, *19*, 486–489. [CrossRef]
15. Li, H.; Pei, W.; Huang, W.; Wang, M.; Wang, Z. Highly Efficient Nanosecond 1.7 μm Fiber Gas Raman Laser by H_2-Filled Hollow-Core Photonic Crystal Fibers. *Crystals* **2020**, *11*, 32. [CrossRef]
16. Sun, Q.; Mao, Q.; Liu, E.; Rao, R.; Ming, H. *Hollow-Core Photonic Crystal Fiber High-Pressure Gas Cell*; Sampson, D.D., Ed.; SPIE-International Society of Optical Engineering: Perth, WA, Australia, 2008; p. 700455.
17. Bischel, W.K.; Dyer, M.J. Wavelength Dependence of the Absolute Raman Gain Coefficient for the Q(1) Transition in H_2. *J. Opt. Soc. Am. B* **1986**, *3*, 677–682. [CrossRef]
18. Xiao, L.; Demokan, M.S.; Jin, W.; Wang, Y.; Zhao, C.-L. Fusion Splicing Photonic Crystal Fibers and Conventional Single-Mode Fibers: Microhole Collapse Effect. *J. Lightwave Technol.* **2007**, *25*, 3563–3574. [CrossRef]
19. Hanson, F.; Poirier, P. Stimulated Rotational Raman Conversion in H_2, D_2, and HD. *IEEE J. Quantum Electron.* **1993**, *29*, 2342–2345. [CrossRef]
20. Herring, G.C.; Dyer, M.J.; Bischel, W.K. Temperature and Density Dependence of the Linewidths and Line Shifts of the Rotational Raman Lines in N_2 and H_2. *Phys. Rev. A* **1986**, *34*, 1944–1951. [CrossRef] [PubMed]

Article

Geometrical Scaling of Antiresonant Hollow-Core Fibers for Mid-Infrared Beam Delivery

Ang Deng and Wonkeun Chang *

School of Electrical and Electronic Engineering, Nanyang Technological University, Singapore 639798, Singapore; deng0082@e.ntu.edu.sg
* Correspondence: wonkeun.chang@ntu.edu.sg

Abstract: We numerically investigate the effect of scaling two key structural parameters in antiresonant hollow-core fibers—dielectric wall thickness of the cladding elements and core size—in view of low-loss mid-infrared beam delivery. We demonstrate that there exists an additional resonance-like loss peak in the long-wavelength limit of the first transmission band in antiresonant hollow-core fibers. We also find that the confinement loss in tubular-type hollow-core fibers depends strongly on the core size, where the degree of the dependence varies with the cladding tube size. The loss scales with the core diameter to the power of approximately −5.4 for commonly used tubular-type hollow-core fiber designs.

Keywords: hollow core fiber; optical waveguide design; mid-infrared beam delivery

Citation: Deng, A.; Chang, W. Geometrical Scaling of Antiresonant Hollow-Core Fibers for Mid-Infrared Beam Delivery. *Crystals* **2021**, *11*, 420. https://doi.org/10.3390/cryst11040420

Academic Editors: David Novoa and Nicolas Y. Joly

Received: 3 March 2021
Accepted: 11 April 2021
Published: 14 April 2021

1. Introduction

Antiresonant hollow core fibers (AR-HCFs) have attracted significant attention from researchers in optics community in recent years. The ability to guide light in the hollow region with only small overlap with the surrounding dielectric material is their main appeal [1–3]. It enables guided optics without many of the limitations imposed by intrinsic optical properties of the waveguide material. In this regard, guiding mid-infrared (mid-IR) beam is one of the key potential applications of AR-HCFs [4–7]. Being guided in the hollow region, the core modes in AR-HCFs can bypass high absorption loss in silica in mid-IR and permit fiber guidance beyond 2.4 μm wavelength [8,9]. For example, Ref. [10] reports AR-HCF transmission loss of 18 dB·km^{-1} at 3.1 μm and 40 dB·km^{-1} at 4 μm. Ref. [11] demonstrates low-loss guidance over a broad mid-IR spectral band spanning 2.5–7.9 μm.

For low-loss mid-IR operation, the cross-sectional geometry of AR-HCF needs to be scaled up from those that guide near-infrared or visible. This is because the waveguide effect remains roughly the same when both the geometry and wavelength are scaled by the same amount [12,13]. Our question is then how the guidance in mid-IR is influenced by the scaling of individual structural parameters. There are two key fiber parameters that are crucial in this respect. The dielectric cladding wall thickness is one of them. It dictates the spectral limits of the transmission bands in AR-HCF. However, its effect on the guidance in the long-wavelength limit of the first transmission band, which is the region most relevant to its mid-IR operation, has not yet been discussed in detail. The core size is the other parameter that has tremendous effect on mid-IR guidance. This is clearly reflected in past mid-IR-guiding AR-HCF demonstrations, where the core diameters are usually much larger than those designed for near-infrared or visible regions, measuring up to 119 μm [11].

In this work, we investigate numerically the geometrical scaling of AR-HCFs and its influence on the confinement loss. We focus on studying the loss in the long-wavelength limit and identify factors that need to be considered when designing low-loss AR-HCFs for mid-IR applications. As we shall see, the wall thickness must be kept thicker than approximately a quarter of the wavelength to achieve low-loss guidance in mid-IR. Moreover, our results show that the confinement loss in tubular-type AR-HCF becomes more strongly dependent on the core size when the cladding tube size is optimized for low loss.

2. Design and Its Geometrical Formalism

One of the most successful branches of AR-HCFs is the tubular type where the central hollow core is surrounded by a small number of thin-wall tubular cladding elements [5]. The tubular hollow-core fiber (THCF) with its simple geometry is relatively easy to fabricate, yet it can still exhibit transmission loss as low as just a few tens of dB·km^{-1} [10,14]. An idealized cross-sectional structure of a six-element THCF is presented in Figure 1a, where gray shades correspond to the regions of dielectric medium, and white regions are hollow. D is the core diameter which is defined as the diameter of the largest circle that fits in the middle hollow region; d is the exterior diameter of the tubular cladding elements; and t is their wall thickness. The size of the tubular cladding elements in THCF is often characterized in terms of the ratio d/D. The range of possible d/D is then set by the number of cladding elements, N, i.e., $0 < d/D < \sin(\pi/N)/(1 - \sin(\pi/N))$ [15].

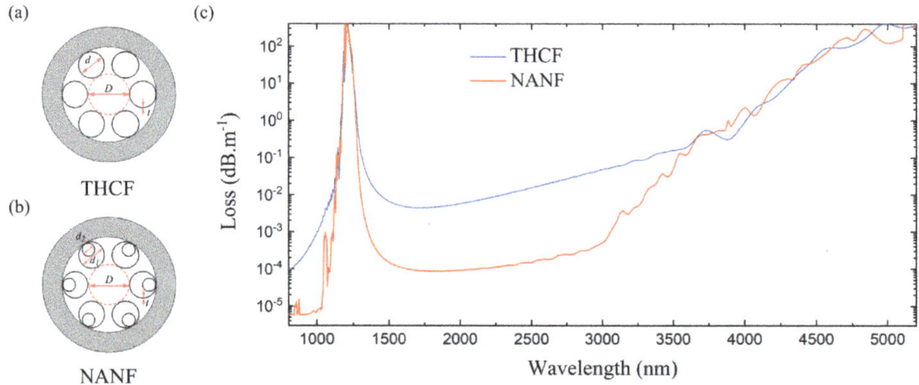

Figure 1. Idealized cross-sections of (**a**) a six-element tubular hollow-core fiber (THCF) and (**b**) a six-element nested antiresonant nodeless fiber (NANF). Gray shades correspond to the regions of dielectric medium, and white regions are hollow. D is the core diameter which is defined as the diameter of the largest circle that fits in the central hollow region. t is the dielectric wall thickness of the cladding elements. d is the exterior diameter of the cladding elements in THCF, and d_1 and d_2 are those of the outer and inner cladding tubes, respectively, in NANF. (**c**) Confinement loss of silica-based six-element THCF and NANF calculated using finite-element method. The structural parameters are D = 60 µm, t = 0.6 µm, and $d = d_1 = 2d_2 = 39$ µm.

A variant of THCF, called nested antiresonant nodeless fiber (NANF), has additional smaller tubular elements of the same wall thickness nested inside the original cladding tubes of THCF, which provide one more antiresonant reflection layer to enhance the light confinement [16,17]. Their presence substantially reduces the loss, and NANFs that exhibit ultra-low loss—comparable to the loss in telecommunication fibers—have been reported in recent years [18,19]. Figure 1b is an idealized structure of a six-element NANF. In this design, the cladding elements are characterized by the exterior diameter of the outer cladding tubes, d_1, and that of the inner tubes, d_2. The core size, D, and dielectric wall thickness, t, are defined in the same way as in THCF.

Figure 1c presents the confinement loss of the fundamental core mode in silica-based THCF and NANF calculated using finite-element modeling. We apply a 20 µm thick perfectly matched layer at the boundary of our calculation domain. We set D = 60 µm, t = 0.6 µm, and $d = d_1 = 2d_2 = 39$ µm in both fibers. Their loss spectra consist of low and high loss regions, where in the low-loss regions, NANF outperforms THCF by approximately two orders of magnitude. The locations of the high-loss bands in AR-HCFs are dictated by the dielectric wall thickness of cladding elements. This gives rise to presence of the peak seen at around 1.25 μm in both geometries. At this wavelength, dielectric wall essentially acts as a Fabry–Perot resonator, and light can easily escape from the core resulting in leakage loss. The resonant wavelengths are given by [20,21]:

$$\lambda_m = \frac{2t}{m}\sqrt{n^2 - 1} \tag{1}$$

where t is the wall thickness, and n is the refractive index of the dielectric material, silica. m is a positive integer which represents the order of the resonance. Hence, the loss at 1.25 μm is the first resonant band, i.e., $m = 1$, and the low-loss region on the long-wavelength side of it is the first transmission band.

Two most influential structural parameters that govern the light guidance in AR-HCF are the dielectric wall thickness of the cladding elements and core diameter. We can express them in terms of the operating wavelength to generalize our investigation for all wavelengths, and study individually effect of each on the confinement loss. For this, we first introduce normalized frequency, F, which represents dielectric wall thickness relative to the wavelength. This is given by:

$$F = \frac{2t}{\lambda}\sqrt{n^2 - 1} \tag{2}$$

Where λ is the wavelength. We assume for now the refractive index of silica, n, is fixed at 1.45 in all calculations to follow. The consequence of its large variation near the silica material resonance at around 8 μm shall be discussed in the later part in Section 3. In Equation (2), F naturally follows the high-loss and low-loss bands in AR-HCF. That is, F close to an integer value ($F \approx m$) corresponds to a resonant band while F close to a half integer value ($F \approx m - 0.5$) implies a transmission band, where $m = 1, 2, 3, \ldots$ is the order of the resonant or transmission band. Next, the core diameter, D, can also be standardized and expressed in terms of λ by introducing the normalized core diameter, G, which is given by:

$$G = \frac{D}{\lambda} \tag{3}$$

Our comprehensive literature survey indicates that G in the range of 25–35 is used the most often, as it strikes a good balance between the confinement and bending-induced loss [1,13].

It is important to note when the entire cross-sectional structure of AR-HCF and wavelength are scaled by the same factor, the imaginary part of effective indices of the core modes—responsible for the confinement loss in respective modes—remain the same. We corroborate this assertion in Figure 2a with a plot of the imaginary part of effective index of the fundamental core mode, $\Im(n_{\text{eff}})$, in a six-element THCF versus the scaling factor, f. Here, the calculations are carried out for the wavelength, $f\lambda$, on the fiber geometry characterized by fD, fd, and ft as illustrated in Figure 2b. Namely, the parameters used in the calculation are $\lambda = 1.06$ μm, $D = 26.5$ μm, $d = 17.2$ μm, and $t = 0.26$ μm. The confinement loss, α_{dB}, expressed in dB·m^{-1} and $\Im(n_{\text{eff}})$ are related through [22]:

$$\alpha_{\text{dB}} = 20\log_{10}\left(\exp\left(\frac{2\pi}{\lambda}\Im(n_{\text{eff}})\right)\right) \approx 8.686\left(\frac{2\pi}{\lambda}\Im(n_{\text{eff}})\right) \tag{4}$$

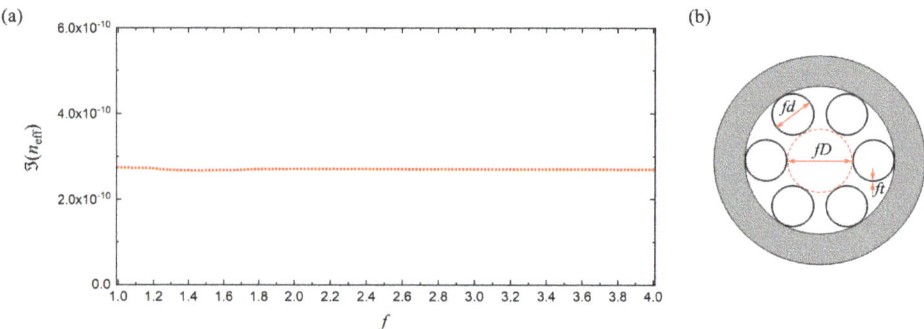

Figure 2. (**a**) Imaginary part of effective index of the fundamental core mode, $\Im(n_{\text{eff}})$, in a six-tube THCF versus f when the fiber structure depicted in (**b**) and wavelength are both scaled by f. The parameters used in the calculation are λ = 1.06 μm, D = 26.5 μm, d = 17.2 μm, and t = 0.26 μm.

Figure 2 and Equation (4) elucidate that α_{dB} is not conserved when the fiber geometry and wavelength are scaled by the same amount. Rather, α_{dB} varies inversely with λ since $\Im(n_{\text{eff}})$ is the quantity that is conserved.

3. Results and Discussions

Figure 3a presents the imaginary part of the effective index of the fundamental core mode, $\Im(n_{\text{eff}})$, as a function of the normalized frequency, F, in a six-tube THCF when G = 25 and d/D = 0.65. Note d/D = 0.65 is the ratio that gives the lowest loss in a six-tube THCF [15]. We illustrate the excellent scalability of our study by repeating the calculations for four different values of t = 0.4, 0.6, 0.8, and 1 μm. The results are accurately reproduced in all four calculations, including the highly oscillatory behaviors in the high-loss regions in Figure 3a. Thus, F allows us to carry out the study in the most general manner.

The presence of the transmission and resonant bands can be seen clearly in Figure 3a. The loss peaks are shifted slightly from the integer values and placed at $F \approx m + 0.1$, while the lowest $\Im(n_{\text{eff}})$ in the mth-order transmission band appears at $F \approx m - 0.5$ indicated with dashed-vertical lines. The small shift of the loss peak from the integer F values has also been noted in several past numerical studies [3,23]. A comprehensive encounter of the origin behind the shift is presented in Reference [23]. We note that the widths and lowest $\Im(n_{\text{eff}})$ values are roughly the same across all transmission bands. Moving in the direction of decreasing F from the lowest $\Im(n_{\text{eff}})$ point in each transmission band, we begin to observe an oscillation and rise in the loss at $F \approx m - 0.6$. They appear in remarkably similar manner in all transmission bands with distinctive small peaks that are located near the red-edges indicated by dotted-vertical lines. The high loss regions, or the mth-order resonant bands, are in $m - 0.1 < F < m + 0.2$. The loss gradually increases towards the band edges of transmission bands. As F approaches a resonance, the index mismatch between the fundamental core mode and dielectric cladding modes decreases, while their spatial field overlap increases, resulting in strong coupling between them. This leads to large light leakage in the core and corresponding rise in the confinement loss [23,24].

To give an idea on how Figure 3a translates to the actual loss in dB·m^{-1} and wavelength, we plot, in Figure 3b, the corresponding confinement loss in the case of two different t values, 0.6 and 0.8 μm, with the equivalent wavelengths for the two cases shown in the upper horizontal axis. For long wavelength operation, light guidance in the first transmission band, i.e., $F < 1$, is of high relevance. In particular, the confinement loss at small F values—in the gray-shaded regions in Figure 3a,b—reveals how THCF is expected to operate in mid-IR. For instance, $F < 0.4$ corresponds to the wavelength longer than 3.15 μm when t = 0.6 μm. One striking observation which we discuss for the first time is how the resonance-like loss starts to show up below $F \approx 0.4$, in a much similar fashion to how it happens around the resonances at $F = 1$, 2, and so on. In other words, after the minimum confinement loss is reached at $F = 0.5$, the loss increases again as if there is another t-induced resonance located at $F = 0$. Even the small peaks at the red-edges of the transmission bands, marked with vertical-dotted lines at $F = 2.26$ and 1.23, are replicated in the first transmission band at $F = 0.21$. Considering that F corresponds to the number of half wavelengths that fits in the thickness of the dielectric cladding tube wall, it is interesting to note that a Fabry–Perot-like resonance still exists when the thickness of the dielectric wall approaches zero. This suggests that the dielectric wall thickness in THCF must be increased accordingly to ensure $F > 0.2$ and achieve low-loss guidance in the long wavelength limit.

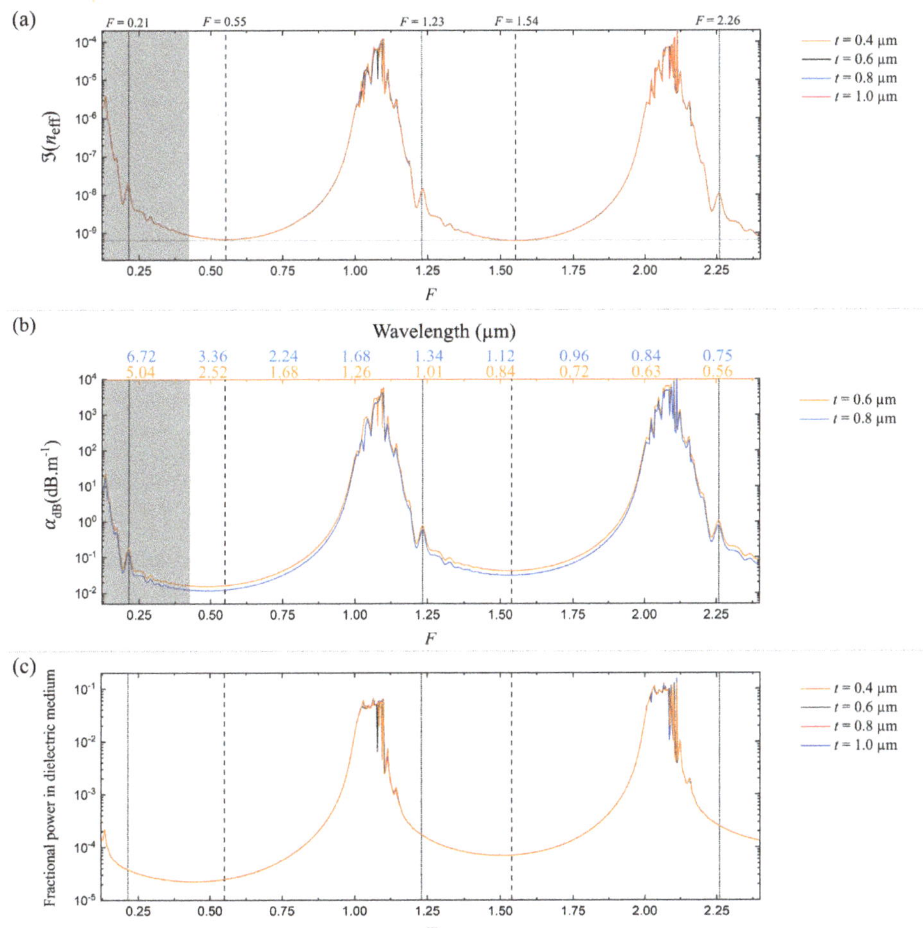

Figure 3. (**a**) Imaginary part of the fundamental core mode index, $\Im(n_{\mathrm{eff}})$, as a function of F in a six-tube THCF when $G = 25$ and $d/D = 0.65$. The calculation is carried out for four different t values, 0.4, 0.6, 0.8, and 1 μm, to illustrate the scalability of the results. (**b**) Corresponding confinement loss versus wavelengths when $t = 0.6$ and 0.8 μm. Gray-shaded regions in (**a**,**b**) mark the long wavelength region, e.g., wavelength longer than 3.15 μm when $t = 0.6$ μm. (**c**) Fractional optical power in the dielectric medium in the same THCF.

Figure 3c shows the corresponding fractional optical power that overlaps with the dielectric medium in the fiber. Its qualitative features are identical to the loss. In the resonant bands, a large portion of light leaks out into the dielectric medium, while overlap is very small in the transmission bands. The overlap becomes smaller as the order of the transmission band is decreased. This quantity is among the most decisive factors that determine the performance of mid-IR guidance in silica-based AR-HCFs. A fiber design that ensures minimum fractional power in silica must be chosen to mitigate the dramatic increase in silica's absorption beyond 2.4 μm. Figure 3c implies that this can be achieved by placing the mid-IR wavelength near $F = 0.5$.

We use the same approach and calculate $\Im(n_{\text{eff}})$, α_{dB}, and fractional optical power in the dielectric medium in a six-element NANF as a function of F. These are shown in Figure 4a–c, respectively. The structural parameters are kept the same as in Figure 3, i.e., $G = 25$, $d_1/D = 0.65$. The ratio between the exterior diameters of the outer and inner cladding elements, $d_1/d_2 = 2$ is used in all calculations. In Figure 4b, we set $t = 0.6$ μm to evaluate the corresponding confinement loss and wavelength. The same quantities for THCF, identical to those presented in Figure 3, are plotted together in Figure 4 for comparison. The reduction of over two orders of magnitude is evident in the transmission bands, clearly demonstrating its superior performance. Towards both edges of each transmission band in NANF, the loss rises rapidly and reaches the same level as in THCF beyond the vertical-dotted lines. This happens when $F < m - 0.75$ in the long- and $F > m$ in the short-wavelength edges. On the other hand, NANF does not further reduce the light-glass overlap as shown in Figure 4c. Therefore, NANF does not offer additional benefit when it comes to mitigating the high material absorption in mid-IR. Furthermore, NANF also exhibits a resonance-like loss in the long-wavelength side of its fundamental transmission band, i.e., as F approaches 0, just like that seen in THCF in Figure 3. This appears to be a feature that is universal to all AR-HCFs.

Thus far, we adopted D correspondingly as F is varied such that $G = 25$ is maintained, which eliminates the effect of core size on the confinement loss in our study. Let us now look at how the core size, D, influences the mid-IR guidance in AR-HCFs. Generally, the loss decreases with increasing core size due to larger glancing angle of light incident at the core boundary, which results in higher Fresnel reflection reducing the attenuation [1]. We start from the Marcatili and Schmeltzer model which presents an analytic expression for the mode attenuation in an idealized capillary that consists of a circular core surrounded by infinitely thick homogenous non-absorbing dielectric medium. In this model, the imaginary part of the fundamental core mode index can be expressed as [22]:

$$\Im(n_{\text{eff}}) = \left(\frac{\lambda}{\pi D}\right)^3 u^2 \frac{n^2+1}{2\sqrt{n^2-1}} \tag{5}$$

Here, $u = 2.405$ is the first zero of the Bessell function J_0. The loss per unit length in a dielectric capillary can then be obtained through Equation (4). It indicates that the confinement loss in a dielectric capillary scales with the inverse cube of D. For a thin wall dielectric capillary, Zeisberger and Schmidt analytically showed that the loss becomes proportional to D^{-4} [25]. This scaling changes to $D^{-4.5}$ in THCF when the tubular cladding elements are in contact with each other, i.e., $d/D = \sin(\pi/N)/(1-\sin(\pi/N))$, as demonstrated numerically in Reference [26]. In this regard, we highlight that the two of the cases discussed above—the capillary with infinitely thick dielectric medium and THCF with touching cladding elements—represent two extreme values of d/D in THCF. On one end, we have the dielectric capillary when $d/D = 0$ where the loss scales with D to the power -3, while on the other end, we arrive at THCF with touching cladding elements when d/D is at its maximum where the loss scales with D to the power -4.5. An interesting question is then how this scaling changes between the two limiting cases. Knowing that the confinement loss of a six-element THCF remains relatively flat and low when $0.5 < d/D < 0.8$ [27,28], understanding how the loss scales with the core size in this range is of particular interest.

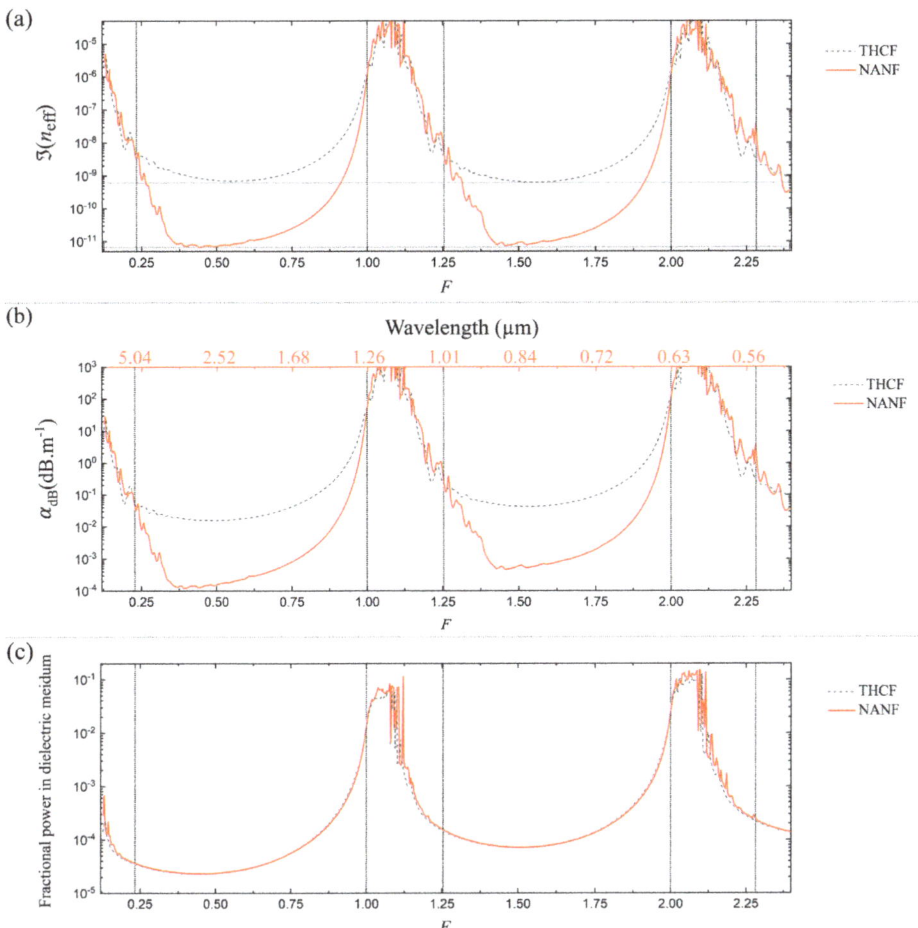

Figure 4. (**a**) Imaginary part of the fundamental core mode index, $\Im(n_{\mathrm{eff}})$, as a function of F in a six-tube NANF when $G = 25$ and $d_1/D = 0.65$, and $d_1 = 2d_2$. (**b**) Corresponding confinement loss versus wavelength when $t = 0.6$ μm. (**c**) Fractional optical power in the dielectric medium in the same NANF. The same quantities for THCF with the identical geometrical parameters are plotted in gray-dashed lines for comparison.

Figure 5 presents our comprehensive encounter of the core size investigation. We calculate $\Im(n_{\mathrm{eff}})$ in THCF as a function of G. Other structural parameters that influence the guidance are varied one-by-one independently to give a clear overview of how the guidance changes with the core size under different conditions. First, we consider the case where N and d/D are fixed at 6 and 0.65, respectively, and calculate the loss as a function of G for five different values of F, i.e., $F = 0.4$, 0.5, 0.6, and 0.8 in the first transmission band and 1.5 in the second transmission band. The results are plotted in Figure 5a. It shows that for all F considered, the loss reduces with increase in G at almost the same rate. By fitting the data numerically using least squares method, we work out that $\Im(n_{\mathrm{eff}})$ is proportional to $G^{-5.4}$. This translates to the confinement loss being proportional to $\lambda^{4.4}/D^{5.4}$ via Equation (4). Similarly, the changes in $\Im(n_{\mathrm{eff}})$ as a function of G for THCF with different number of cladding elements, while fixing $F = 0.5$, i.e., at the center of the first transmission band, are presented in Figure 5b. For each N, the cladding tube size that exhibits low loss, as determined in Reference [15], is used in the calculations. These are $d/D = 0.79$, 0.65, 0.6, 0.5,

0.45 for $N = 5, 6, 7, 8, 9$, respectively. We find that the rate of reduction against G does not vary much with change in N when d/D that gives low loss is chosen in each N. In fact, we obtain the same loss decay rate of 5.4 from Figure 5b. This demonstrates that the dielectric wall thickness and number of cladding elements do not have significant impact on the rate of reduction in $\Im(n_{\text{eff}})$ with increase in the core size.

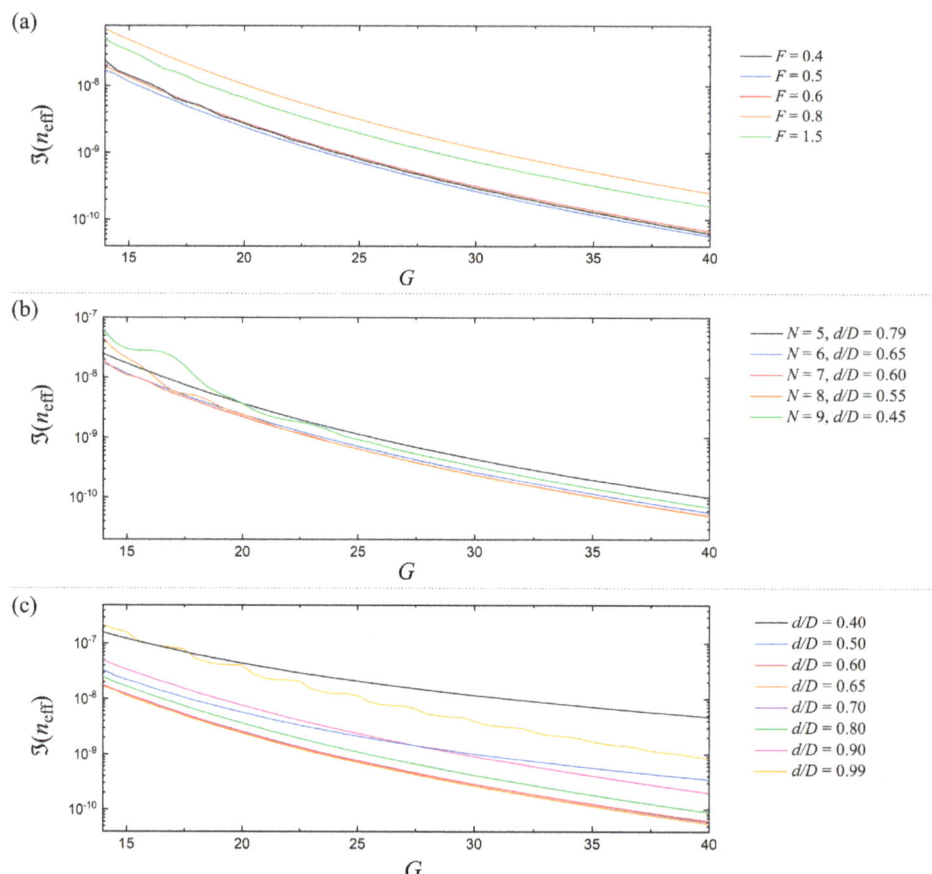

Figure 5. $\Im(n_{\text{eff}})$ versus G in THCF for different structural parameters. (**a**) For $F = 0.4, 0.5, 0.6, 0.8, 1.5$, while $N = 6$ and $d/D = 0.65$. (**b**) For $N = 5, 6, 7, 8, 9$, while $F = 0.5$ and d/D is set at a value that gives low loss in each N as identified in Reference [15]. (**c**) For $d/D = 0.4, 0.5, 0.6, 0.65, 0.7, 0.8, 0.9, 0.99$, while $N = 6$ and $F = 0.5$.

On the contrary, it turns out that the size of the cladding tubes has a major impact. Figure 5c shows the change in $\Im(n_{\text{eff}})$ against G for different sizes of the cladding tubes, while F and N are set at 0.5 and 6. The rate of reduction varies significantly depending on the size of the cladding tubes. This is in accordance with the fact that the two limiting cases of the d/D in THCF, i.e., $d/D = 0$ and $d/D = \sin(\pi/N)/(1 - \sin(\pi/N))$ have largely different rates, G^{-3} for the former and $G^{-4.5}$ for the latter.

Having identified that cladding tube size is the key factor affecting the rate of change of the loss when the core size is varied, we now study in detail how the rate changes against the cladding tube size. Figure 6 reveals the decay rate, x, of $\Im(n_{\text{eff}})$ on G versus the standardized cladding size, $(d/D)'$, when $N = 6, 7, 8$, and $F = 0.5$. Here, the decay rate, x, means that $\Im(n_{\text{eff}})$ is proportional to G^{-x}, which implies through Equation (4) that α_{dB} is proportional to λ^{x-1}/D^x. At each data point in Figure 6, x is numerically calculated by fitting the $\Im(n_{\text{eff}})$ versus G plot using the least squares method in the range $20 < G < 40$. Moreover, since the possible range of d/D in THCF depends on N, i.e., from 0 to $(\sin(\pi/N))/(1-\sin(\pi/N))$, we introduce a new parameter $(d/D)'$ to standardize the cladding size for all N for easy comparison. This is defined as:

$$(d/D)\prime = \frac{d/D}{(\sin(\pi/N))/(1-\sin(\pi/N))} = \frac{d}{D}(\csc(\pi/N)-1) \quad (6)$$

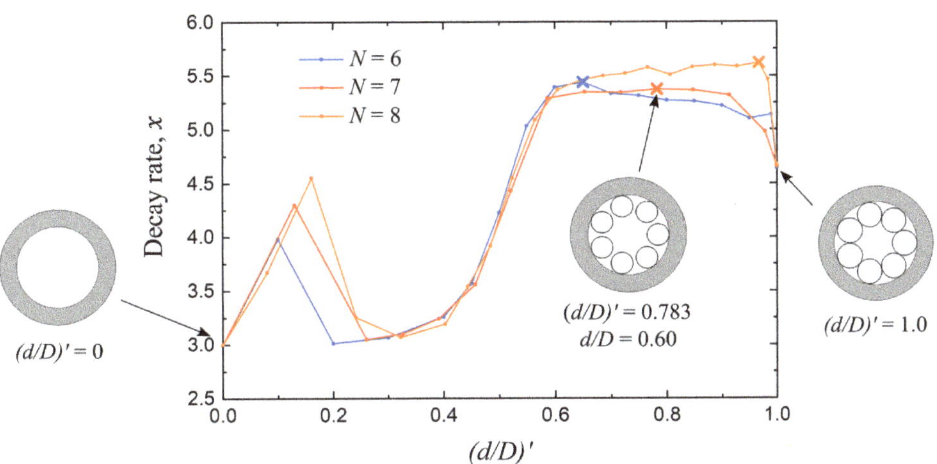

Figure 6. Confinement loss decay rate, x, versus the standardized cladding size, $(d/D)'$, for THCF when $N = 6, 7, 8$, and $F = 0.5$. The decay rate, x, means α_{dB} is proportional to λ^{x-1}/D^x. At each data point, x is calculated by fitting the $\Im(n_{\text{eff}})$ versus G plot using the least squares method in the range $20 < G < 40$. The standardized cladding size, $(d/D)'$, is obtain by normalizing d/D to its respective maximum, $\sin(\pi/N)/(1-\sin(\pi/N))$. Example geometries at three different $(d/D)'$ values when $N = 7$ are illustrated in the insets for reference. The colored crosses mark $(d/D)'$ values that give the lowest loss in each N which coincides with the largest decay rate.

The lines have remarkably similar shapes for all N. The decay rate does not vary monotonically with d/D. They all start at 3 when $(d/D)' = 0$, and rise to values between 4 and 4.5 when $(d/D)'$ is around 0.15–0.2. Upon further increase in $(d/D)'$, they quickly dip and remain at around 3 in the range $0.2 < (d/D)' < 0.4$. Beyond 0.4, the rates rise and plateau in between 5 and 5.5. This is the range most relevant as majority of THCFs have $(d/D)'$ of 0.6–0.9 [15]. They decrease rapidly near the maximum $(d/D)'$ of 1, arriving at 4.5. We believe this non-monotonic variation of x is mainly due to the presence or lack thereof airy modes in the space in-between the cladding tubes depending on $(d/D)'$, as well as the influence of the large bulk of dielectric material in the outer jacket tube that grows as $(d/D)'$ decreases. One striking observation in Figure 6 is that the largest decay rate in each N coincides with $(d/D)'$ that gives the lowest loss. These are indicated with the crosses in Figure 6. In other words, the confinement loss in the best performing THCF has the strongest dependence on the core size.

From the results presented so far, it is made clear that the dielectric wall thickness must scale proportionately with the wavelength such that $F \approx 0.5$ is maintained even in the mid-IR region. Moreover, the core size must be sufficiently large and carefully chosen since the confinement loss increases proportionally to approximately $\lambda^{4.4}/D^{5.4}$ in low-loss THCFs. However, there is another important design aspect which we have not brought up in this work, i.e., the index of the dielectric material. This also needs to be carefully considered for realizing low-loss mid-IR guidance in AR-HCFs. One interesting aspect in this regard is that the refractive index of silica can drop below 1 in mid-IR at around 7.3 μm [29]. While we can imagine that this may lead to a unique situation where the hollow-core fiber exhibits index guidance, its performance will be severely restricted by the high material absorption in this spectral region [9,11,29]. Therefore, other compound-glass materials that are transparent in this range, such as chalcogenide glass, should be considered as the waveguide material for guidance at around 8 μm wavelength.

4. Conclusions

In this work, we present comprehensive numerical study on the effect of scaling two key structural parameters of AR-HCFs—dielectric wall thickness of the cladding elements and core diameter—in view of low-loss mid-IR beam delivery. We show for the first time the appearance of resonance-like loss peak in the long-wavelength limit of the first transmission band in AR-HCF. This indicates that the dielectric wall thickness of the cladding elements must be increased to at least a quarter of the wavelength to ensure low-loss guidance in the long wavelength region. Furthermore, we reveal that the rate of loss reduction with increase in the core diameter in THCF depends on the cladding tube size. The loss depends most strongly on the core size when the cladding tube diameter is optimized for low loss. It increases proportionally to approximately $\lambda^{4.4}/D^{5.4}$ when $(d/D)'$ is in the range 0.6–0.96.

Author Contributions: Conceptualization, W.C.; Data curation, A.D.; Formal analysis, A.D. and W.C.; Funding acquisition, W.C.; Investigation, A.D. and W.C.; Methodology, A.D.; Project administration, W.C.; Supervision, W.C.; Validation, A.D. and W.C.; Visualization, A.D.; Writing—original draft, A.D.; Writing—review & editing, A.D. and W.C. All authors have read and agreed to the published version of the manuscript.

Funding: National Research Foundation—Singapore (QEP-P4).

Institutional Review Board Statement: Not applicable.

Informed Consent Statement: Not applicable.

Data Availability Statement: Not applicable.

Acknowledgments: The authors acknowledge helpful discussion with Muhammad Rosdi Abu Hassan.

Conflicts of Interest: The authors declare no conflict of interest.

References

1. Yu, F.; Knight, J.C. Negative curvature hollow-core optical fiber. *IEEE J. Sel. Top. Quantum Electron.* **2016**, *22*, 146–155. [CrossRef]
2. Wei, C.; Weiblen, R.J.; Menyuk, C.R.; Hu, J. Negative curvature fibers. *Adv. Opt. Photonics* **2017**, *9*, 504–561. [CrossRef]
3. Vincetti, L.; Setti, V. Waveguiding mechanism in tube lattice fibers. *Opt. Express* **2010**, *18*, 23133–23146. [CrossRef]
4. Harrington, J.A. A review of IR transmitting, hollow waveguides. *Fiber Integr. Opt.* **2000**, *19*, 211–227. [CrossRef]
5. Pryamikov, A.D.; Biriukov, A.S.; Kosolapov, A.F.; Plotnichenko, V.G.; Semjonov, S.L.; Dianov, E.M. Demonstration of a waveguide regime for a silica hollow-core microstructured optical fiber with a negative curvature of the core boundary in the spectral region >3.5 μm. *Opt. Express* **2011**, *19*, 1441–1448. [CrossRef]
6. Hassan, M.R.A.; Yu, F.; Wadsworth, W.J.; Knight, J.C. Cavity-based mid-IR fiber gas laser pumped by a diode laser. *Optica* **2016**, *3*, 218–221. [CrossRef]
7. Lee, E.; Luo, J.; Sun, B.; Ramalingam, V.; Zhang, Y.; Wang, Q.; Yu, F.; Yu, X. Flexible single-mode delivery of a high-power 2 μm pulsed laser using an antiresonant hollow-core fiber. *Opt. Lett.* **2018**, *43*, 2732–2735. [CrossRef] [PubMed]
8. Humbach, O.; Fabian, H.; Grzesik, U.; Haken, U.; Heitmann, W. Analysis of OH absorption bands in synthetic silica. *J. Non-Cryst. Solids* **1996**, *203*, 19–26. [CrossRef]
9. Wu, D.; Yu, F.; Liao, M. Understanding the material loss of anti-resonant hollow-core fibers. *Opt. Express* **2020**, *28*, 11840–11851. [CrossRef]
10. Yu, F.; Song, P.; Wu, D.; Birks, T.; Bird, D.; Knight, J. Attenuation limit of silica-based hollow-core fiber at mid-IR wavelengths. *Appl. Phys. Lett.* **2019**, *4*, 080803. [CrossRef]
11. Kolyadin, A.N.; Kosolapov, A.F.; Pryamikov, A.D.; Biriukov, A.S.; Plotnichenko, V.G.; Dianov, E.M. Light transmission in negative curvature hollow core fiber in extremely high material loss region. *Opt. Express* **2013**, *21*, 9514–9519. [CrossRef]
12. Vincetti, L. Empirical formulas for calculating loss in hollow core tube lattice fibers. *Opt. Express* **2016**, *24*, 10313–10325. [CrossRef] [PubMed]
13. Yu, F.; Knight, J.C. Spectral attenuation limits of silica hollow core negative curvature fiber. *Opt. Express* **2013**, *21*, 21466–21471. [CrossRef] [PubMed]

14. Yu, F.; Wadsworth, W.J.; Knight, J.C. Low loss silica hollow core fibers for 3–4 μm spectral region. *Opt. Express* **2012**, *20*, 11153–11158. [CrossRef]
15. Deng, A.; Hasan, I.; Wang, Y.; Chang, W. Analyzing mode index mismatch and field overlap for light guidance in negative-curvature fibers. *Opt. Express* **2020**, *28*, 27974–27988. [CrossRef]
16. Poletti, F. Nested antiresonant nodeless hollow core fiber. *Opt. Express* **2014**, *22*, 23807–23828. [CrossRef]
17. Belardi, W.; Knight, J.C. Hollow antiresonant fibers with reduced attenuation. *Opt. Lett.* **2014**, *39*, 1853–1856. [CrossRef]
18. Bradley, T.; Hayes, J.; Chen, Y.; Jasion, G.; Sandoghchi, S.R.; Slavík, R.; Fokoua, E.N.; Bawn, S.; Sakr, H.; Davidson, I. Record low-loss 1.3 dB/km data transmitting antiresonant hollow core fibre. In Proceedings of the 2018 European Conference on Optical Communication (ECOC), Rome, Italy, 23–27 September 2018; pp. 1–3.
19. Jasion, G.T.; Bradley, T.D.; Harrington, K.; Sakr, H.; Chen, Y.; Fokoua, E.N.; Davidson, I.A.; Taranta, A.; Hayes, J.R.; Richardson, D.J. Hollow core NANF with 0.28 dB/km attenuation in the C and L bands. In *Proceedings of Optical Fiber Communication Conference*; Optical Society of America: Washington, DC, USA, 2020; p. Th4B. 4.
20. Litchinitser, N.; Abeeluck, A.; Headley, C.; Eggleton, B. Antiresonant reflecting photonic crystal optical waveguides. *Opt. Lett.* **2002**, *27*, 1592–1594. [CrossRef]
21. Duguay, M.; Kokubun, Y.; Koch, T.L.; Pfeiffer, L. Antiresonant reflecting optical waveguides in SiO_2-Si multilayer structures. *Appl. Phys. Lett.* **1986**, *49*, 13–15. [CrossRef]
22. Marcatili, E.A.; Schmeltzer, R. Hollow metallic and dielectric waveguides for long distance optical transmission and lasers. *Bell Syst. Tech. J.* **1964**, *43*, 1783–1809. [CrossRef]
23. Vincetti, L.; Rosa, L. A simple analytical model for confinement loss estimation in hollow-core Tube Lattice Fibers. *Opt. Express* **2019**, *27*, 5230–5237. [CrossRef] [PubMed]
24. Debord, B.; Amsanpally, A.; Chafer, M.; Baz, A.; Maurel, M.; Blondy, J.-M.; Hugonnot, E.; Scol, F.; Vincetti, L.; Gérôme, F. Ultralow transmission loss in inhibited-coupling guiding hollow fibers. *Optica* **2017**, *4*, 209–217. [CrossRef]
25. Zeisberger, M.; Schmidt, M.A. Analytic model for the complex effective index of the leaky modes of tube-type anti-resonant hollow core fibers. *Sci. Rep.* **2017**, *7*, 1–13. [CrossRef]
26. Masruri, M.; Cucinotta, A.; Vincetti, L. Scaling laws in tube lattice fibers. In Proceedings of the 2015 Conference on Lasers and Electro-Optics (CLEO), San Jose, CA, USA, 10–15 May 2015; pp. 1–2.
27. Song, P.; Phoong, K.Y.; Bird, D. Quantitative analysis of anti-resonance in single-ring, hollow-core fibres. *Opt. Express* **2019**, *27*, 27745–27760. [CrossRef]
28. Belardi, W.; Knight, J.C. Hollow antiresonant fibers with low bending loss. *Opt. Express* **2014**, *22*, 10091–10096. [CrossRef]
29. Kitamura, R.; Pilon, L.; Jonasz, M. Optical constants of silica glass from extreme ultraviolet to far infrared at near room temperature. *Appl. Opt.* **2007**, *46*, 8118–8133. [CrossRef]

Article

Investigation on Chalcogenide Glass Additive Manufacturing for Shaping Mid-infrared Optical Components and Microstructured Optical Fibers

Julie Carcreff [1], François Chevíré [1], Ronan Lebullenger [1], Antoine Gautier [1], Radwan Chahal [2], Jean Luc Adam [1], Laurent Calvez [1], Laurent Brilland [2], Elodie Galdo [1], David Le Coq [1], Gilles Renversez [3] and Johann Troles [1,*]

[1] CNRS, University of Rennes, ISCR-UMR 6226, 35000 Rennes, France; julie.carcreff@univ-rennes1.fr (J.C.); francois.chevire@univ-rennes1.fr (F.C.); ronan.lebullenger@univ-rennes1.fr (R.L.); antoine.gautier@univ-rennes1.fr (A.G.); jean-luc.adam@univ-rennes1.fr (J.L.A.); laurent.calvez@univ-rennes1.fr (L.C.); elodie.galdo@univ-rennes1.fr (E.G.); david.lecoq@univ-rennes1.fr (D.L.C.)

[2] Selenoptics, 263 Avenue Gal Leclerc, 35042 Rennes, France; radwan.chahal@selenoptics.com (R.C.); laurent.brilland@selenoptics.com (L.B.)

[3] CNRS, Centrale Marseille, Institut Fresnel, Aix-Marseille University, UMR 7249, 13013 Marseille, France; gilles.renversez@fresnel.fr

* Correspondence: johann.troles@univ-rennes1.fr; Tel.: +33-(0)2-23-23-67

Citation: Carcreff, J.; Chevíré, F.; Lebullenger, R.; Gautier, A.; Chahal, R.; Adam, J.L.; Calvez, L.; Brilland, L.; Galdo, E.; Le Coq, D.; et al. Investigation on Chalcogenide Glass Additive Manufacturing for Shaping Mid-infrared Optical Components and Microstructured Optical Fibers. *Crystals* **2021**, *11*, 228. https://doi.org/10.3390/cryst11030228

Academic Editor: Shujun Zhang

Received: 6 February 2021
Accepted: 22 February 2021
Published: 25 February 2021

Abstract: In this work, an original way of shaping chalcogenide optical components has been investigated. Thorough evaluation of the properties of chalcogenide glasses before and after 3D printing has been carried out in order to determine the impact of the 3D additive manufacturing process on the material. In order to evaluate the potential of such additive glass manufacturing, several preliminary results obtained with various chalcogenide objects and components, such as cylinders, beads, drawing preforms and sensors, are described and discussed. This innovative 3D printing method opens the way for many applications involving chalcogenide fiber elaboration, but also many other chalcogenide glass optical devices.

Keywords: chalcogenide glasses; 3D printing; mid-infrared fibers; photonic crystal fibers

1. Introduction

In recent years, a growing interest has been developed for optical materials and fibers for the mid-infrared (mid-IR) region, especially for chalcogenide glasses [1,2]. Such interest originates from societal needs for health and environment for instance, and also from the demand for military applications [3,4]. Indeed, the mid-IR spectral region contains the 3–5 μm and 8–12 μm atmospheric transparent windows, where both military and civilian thermal imaging can take place [5,6]. Compared to oxide-based glasses, vitreous materials containing chalcogen elements, i.e., S, Se and Te, show large transparency windows in the infrared. Depending on their chemical composition, chalcogenide glasses can be transparent from the visible up to 12–18 μm [7–9]. In this context, we have investigated an alternative approach based on a 3D printing process for fabricating mid-IR optical components such as preforms, optical fibers, sensors and beads.

3D printing, more formally known as additive manufacturing, was firstly a shaping technique widely used for polymer for rapid prototyping processes but it is now more and more commonly employed for mass production and mass customization, as well [10–13]. Very quickly, this technology was extended to metals [14], ceramics [15] and even glasses [16–18]. The most common method for polymers is the fused filamentation fabrication (FFF) method, also known as fused deposition modeling (FDM) [19,20]. It involves the extrusion of a thermoplastic material through a heated die and accurate deposition, layer by layer, until the desired three-dimensional object is obtained. It has been demonstrated that such filamentation techniques can be modified and applied to the 3D printing of glasses with low glass transition temperature (T_g), also known as "soft glasses", such as chalcogenide glasses [21,22] and phosphate glasses as presented in reference [23]. In previous works, in reference [22], it has been shown that a microstructured optical fiber can be obtained from a 3D printed chalcogenide glass preform. The present study is a more complete investigation of the thermal and optical properties of the chalcogenide glasses after the 3D printing process.

For this work, the 3D-printing set-up is based on a customized desktop RepRap-style 3D printer running Marlin firmware [24]. This printer, typically used with polymer filaments to produce plastic objects, was upgraded to reach a deposition temperature of up to 400 °C required for soft glasses and the feeding mechanism was customized to suitably handle brittle materials such as glasses.

Different physical and optical properties of the printed glasses such as density, thermal expansion coefficient, refractive indices, and transmission have been investigated and compared to the properties of classical melt-quenched glasses. By using this additive manufacturing method, chalcogenide cylinders, pellets, beads, as well as microstructured performs with complex designs can be fabricated in a single step with a high degree of repeatability and an accuracy of the geometry. This original 3D printing method opens the way for numerous applications, involving chalcogenide fiber manufacturing, but also many other chalcogenide glasses optical devices.

2. Materials and Methods

2.1. Thermal and Physical Optical Properties of the Chosen Glass: $Te_{20}As_{30}Se_{50}$

The selected glass for printing shall present thermal properties compatible with manufacture by fused filamentation fabrication. To begin, the use of a commercial extrusion head usually utilized for conventional thermoplastics such as polylactic acid (PLA) requires that the selected glass shows a low glass transition temperature (T_g). In addition, it would be preferable that the glass is stable with respect to crystallization. In this context, the $Te_{20}As_{30}Se_{50}$ (TAS) chalcogenide glass that exhibits a T_g slightly below 140 °C appears to be a good candidate [25]. Figure 1a shows the differential scanning calorimetry (DSC) curve of the TAS glass measured with a DSC Q20 from TA instrument with a heating rate of 10 °C/min. The glass exhibits a T_g close to 137 °C and no sign of crystallization are observed up to 300 °C, which confirms that the glass could be compatible with the FFF 3D printing process.

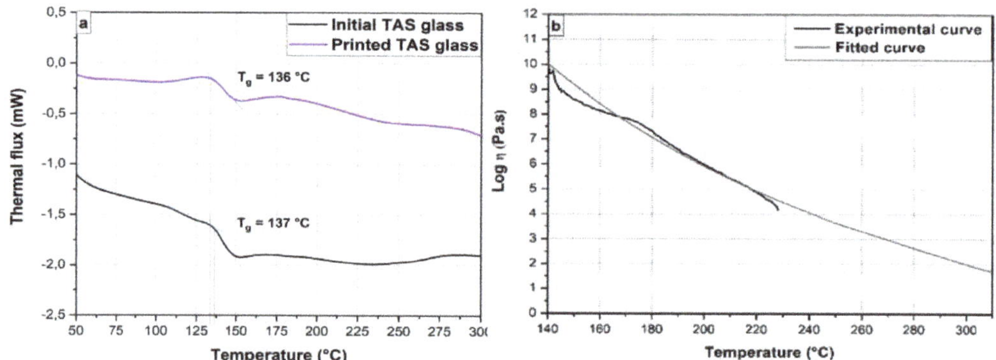

Figure 1. Thermal properties of the $Te_{20}AS_{30}Se_{50}$ (TAS) chalcogenide glasses: (**a**) differential scanning calorimetry measurement, (**b**) viscosity versus temperature curve.

The implementation of a filamentation process with this glass also requires a better knowledge of its viscosity as a function of temperature. For this purpose, a TAS pellet was prepared in order to measure the viscosity. Both sides of the printed disk (15 mm diameter, 4 mm height) were polished to ensure that they were parallel. The viscosity of the glass was measured above the glass transition temperature between 140 and 230 °C by using a Rheotronic®parallel plate viscometer (Theta industries). This experimental set-up did not permit one to measure viscosities lower than 10^4 Pa.s. This value is reached when the temperature of the glass is equal to 230 °C. However, considering that experimental data follow the VFT (Vogel–Fulcher–Tammann) law, the viscosity value can be estimated up to 300 °C using a fitting procedure from Equations (1) and (2) [26], as shown in Figure 1b. This enables one to estimate the flow rate and, therefore to optimize the process.

$$\log_{10} \eta(T) = \log_{10}\eta_\infty + \left(\frac{A}{T - T_0}\right) \tag{1}$$

This equation can also be written as follows:

$$\log_{10} \eta(T) = \log_{10}\eta_\infty + \frac{\left(\log_{10} \eta(T_g) - \log_{10} \eta_\infty\right)^2}{m\left(\frac{T}{T_g} - 1\right) + \left(\log_{10} \eta(T_g) - \log_{10} \eta_\infty\right)} \tag{2}$$

where, $\log_{10}\eta_\infty$, m and T_g are constants determined by the fitting procedure. They are equal to −9, 35.9 and 413.15 K, respectively.

Commonly, the required viscosities for polymers in FFF processes are around 10^2–10^3 Pa.s [27,28]. According to Figure 1b, such viscosities are obtained when the temperature gets to the 270–300 °C range, which confirms that the TAS glass is a suitable candidate for FFF printing at reasonable temperatures.

The optical properties of the glass also have to be studied to ensure its suitability for the targeted infrared optical applications, particularly for the realization of infrared components and fibers.

The transmission of a polished 1-mm thick TAS glass pellet was measured with a Fourier Transform Spectrometer (FTIR) (Bruker Tensor 37) from of 1 µm to 22 µm. The obtained transmission spectrum is shown in Figure 2. The transmission of the glass ranges from 1.2 to 20 µm, with a maximum of transmission of around 65%. Such moderate transmittance is explained by strong Fresnel reflections due to the high refractive index of TAS glass. Indeed, as presented in Figure 2, the refractive index of the TAS glass is close to 3.

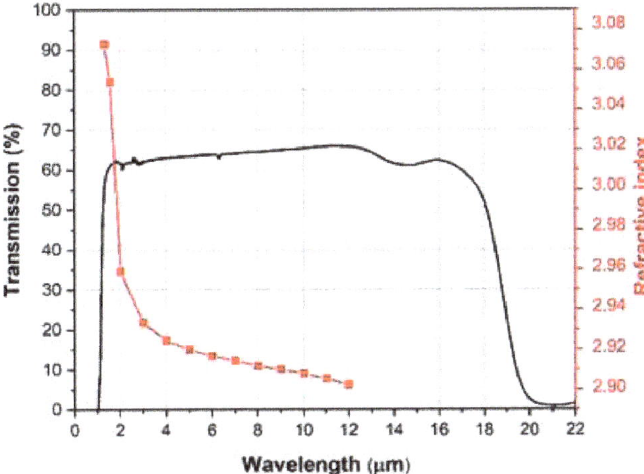

Figure 2. Transmission and refractive indices of $Te_{20}As_{30}Se_{50}$ chalcogenide glass in the mid-infrared spectral region

Refractive indices were measured for different wavelengths between 1.3 and 12 μm. From 2 to 12 μm, a homemade optical bench based on the minimum deviation method or Littrow method [29] was utilized by using a TAS prism [30,31]. Due to the experimental limits of our prism method (2–12 μm), two other measurements have been achieved at 1.31 μm and 1.55 μm with a commercial Metricon device (Model 2010/M). The refractive index of the TAS glass is equal to 3.073 at 1.31 μm and decreases to 2.902 at 12 μm.

2.2. Additive Manufacturing Process

In this study, the 3D-printing set-up is based on a customized commercial RepRap-style 3D printer (Anet A8) upgraded for soft glasses. The feeding mechanism is especially customized for brittle materials. The extrusion head, made of copper, can reach temperatures up to 400 °C. As shown in Figure 2, TAS glass reaches the appropriate viscosity when the temperature is above 270 °C. As a result, the temperature of the printer head was set to around 300 °C. The width of the printed lines resulted from the size of the nozzle diameter, which was mainly chosen at 400 μm. However, we have shown that the printing process can also use a 250 μm diameter nozzle. The chalcogenide glass was deposited on a sodalime silicate glass bed heated to 140 °C, which ensured the good adhesion of the printed sample. It can be noted that all the components printed in this study were extruded using a layer thickness of 100 μm, which corresponds to a z-pitch of 100 μm for each layer.

The 3D printer was fed by 3 mm diameter TAS rods which were prepared from TAS bars synthesized by the melt–quenching method. To save time and to ensure that the printing process was possible with such a glass, the preliminary tests were performed with unpurified TAS glasses. The different chemical elements, i.e., Te (5N), As (5N) and Se (5N), were placed in a silica ampoule under a vacuum. The ampoule was then heated at 850 °C in a rocking furnace for a few hours and finally the molten glass was quenched at room temperature and placed in an annealing furnace at T_g +5 °C for 3 h and then slowly cooled to ambient temperature.

After the validation of the additive manufacturing method for TAS, the raw material rods were made from purified glass using a similar process to the one described in reference [32]. In this case, $TeCl_4$ and Al (1000 ppm and 100 ppm, respectively) were added during the glass synthesis. After the first melt–quench, the glass was distilled for the first time under a dynamic vacuum to remove impurities such as CCl_4 and HCl. The second stage of static vacuum distillation made it possible to remove carbon, silica, alumina and other refractory oxides, for example. The glass was then homogenized at 850 °C for a few hours, then cooled to 550 °C and quenched in water before it was annealed at T_g +5 °C.

Once removed from the synthesis silica tube, the TAS glass bars (also named TAS preforms) were placed in an annular furnace within a homemade fibering tower to be drawn into 3 mm diameter rods. Whether it was an unpurified or purified glass preform, meters of fibers were also drawn in order to measure the optical transmission of the initial glasses that would be used to feed the 3D printer (the details of these characterizations will be presented in Section 3.2.). As an example, a 12 cm long glass bar gives about 2 meters of TAS rod.

The rods were then fed into the printer's head to be extruded. The printer movements were controlled by using G-code programs especially computed for driving displacements compatible with TAS glass. Firstly, as presented in Figure 3, simple objects such as cylinders, disks, and beads were made to study the impact of the 3D printing process on the properties of TAS glass such as density, thermal expansion coefficient, refractive indices and optical attenuation. In a second step, more complex shapes were realized such as microstructured preforms (studied in [22] and Section 3.3) and tapers.

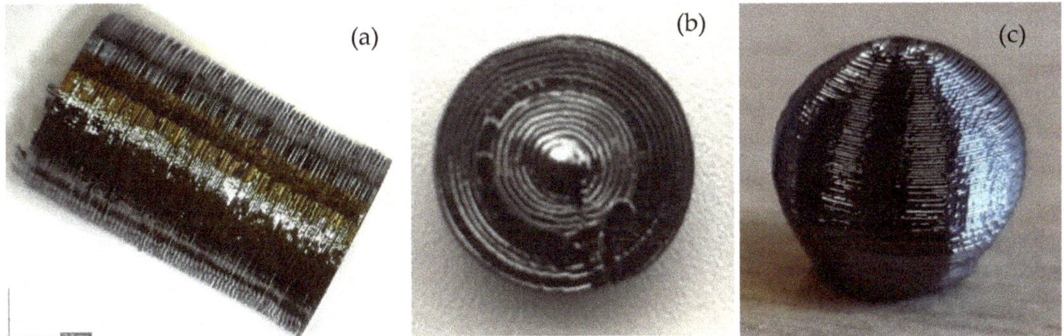

Figure 3. Chalcogenide 3D printed samples: (**a**) chalcogenide cylinder, (**b**) chalcogenide disk (**c**) chalcogenide bead.

3. Results

3.1. Physical Properties of the Printed Glasses

A 5 mm high TAS pellet with a 15 mm diameter was printed as shown in Figure 3b, and polished before being analyzed to compare the physical properties of the initial glass to those of the printed glass. The composition of the pellet was determined by energy dispersive spectroscopy (EDS). The results are shown in Table 1. The comparison with the initial composition of the glass indicates that there is no change in the chemical composition during the printing process. The glass transition temperature (T_g), thermal expansion coefficient (α), density and X-ray diffraction pattern of the printed TAS glass were also compared to a standard TAS glass (see Table 1 and Figure 4). Considering the experimental error inherent to DSC measurements (DSC Q20 Thermal Analysis), the T_g of the printed glass and the initial glass are similar. In addition, no crystallization peak was observed in the DSC measurement of the printed glass (not shown). The thermal coefficient measurements were carried out on a thermomechanical analyzer (TA Instrument, TMA 2940) with a heating ramp of 2 °C·min^{-1}. The coefficient is 22.82·10^{-6} K^{-1} for a classic glass and 22.62·10^{-6} K^{-1} for a printed glass, respectively. Considering the measurement error, both glasses have the same coefficient of expansion. The densities were measured by the Archimedes method on a Mettler Toledo XS64 balance. It is noticed that the density significantly decreases from 4.86 g·cm^{-3} for a quenched glass to 4.81 g·cm^{-3} for a printed glass, meaning that the printed objects do not present the expected density of a bulk glass.

Table 1. Physical property comparison of a melt/quenched glass used as the initial glass and a printed glass.

	Initial Glass	Printed Glass
Composition from EDS analysis (%) (±1)	$Te_{20}As_{30}Se_{50}$	$Te_{21}Se_{29}Te_{50}$
Bulk transmission range (µm)	1–18	1–18
Fiber transmission range (µm)	2–12	2–12
T_g (°C) (±2)	137	136
Coefficient of thermic expansion (10^{-6} K^{-1}) (± 1%)	22.82	22.62
Density (g·cm^{-3}) (± 0.02)	4.86	4.81

Finally, to ensure that the glass did not crystallize during the additive manufacturing process, the initial and the printed TAS pellets were analyzed by X-ray diffraction (XRD). XRD diagrams were recorded at room temperature in the 10–80° 2θ range using a PANalytical X'Pert Pro (Cu $K_{\alpha1}$, $K_{\alpha2}$ radiations, $\lambda_{K\alpha1}$ = 1.54056 Å, $\lambda_{K\alpha2}$ = 1.54439 Å, 40 kV, 40 mA, PIXcel detector 1D). Figure 4 shows that the two diagrams perfectly overlap and no additional diffraction peak is detected for the printed glass.

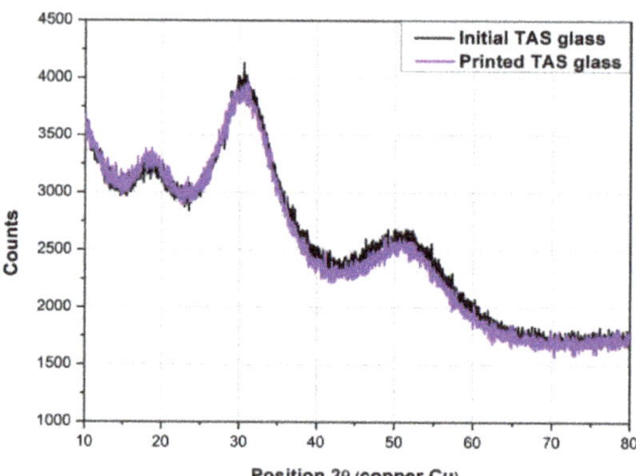

Figure 4. XRD diagrams of initial TAS glass (black curve) and printed TAS glass (purple curve).

3.2. Optical Properties of the Printed Glasses

In order to study the consequences of 3D printing on optical properties, 10 mm diameter chalcogenide beads were 3D printed using concentric extruded layers as illustrated in Figure 5a. Two beads were polished to obtain 1 mm-thick pellets, one along the XY plane and the other one along the XZ plane. The latter, from the process aspect, is equivalent to the plan YZ. It is important to remember that the Z axis is the axis of the layer stacking. The transmission windows of the pellets recorded on a Fourier Transform InfraRed Spectrometer (FTIR, Bruker Vector 22), in the 1.25–22 μm range are shown in Figure 5b. The transmission spectra of the printed and initial TAS glasses are similar in terms of spectral range. However, the maximum transmission of the extruded TAS is limited to 48% for the XY plane and 39% for the XZ plane, compared to 65% for the initial glass.

To observe the effect of the successive addition of the different layers and lines, images were recorded with an infrared camera working in the 7–13 μm wavelength windows (FLIR A655sc). The black frame photograph in Figure 5c corresponds to a TAS pellet made from a melt–quenched glass (used as the initial glass for printing), which seems to be perfectly homogeneous in the infrared range from 7 to 13 μm. The other two photographs are those of the glass pellets printed either in the XZ plane for the blue frame or in the XY plane for the red frame. Imperfections are present in both pads. For the XZ plane, it looks like a periodic accumulation of points attributed to the cross-section of the printed lines while for the XY plane the defects are placed in a circular way corresponding to the in-plane printed lines.

After pellets and beads, two cylinders (or preforms) with an outer diameter of 8 mm were printed. One of the cylinders was obtained from an unpurified TAS glass rod while the second one was obtained from a purified glass rod. These preforms were prepared in order to be drawn and optically characterized with the aim to compare the optical losses between initial glass fibers and printed glass fibers depending on the quality of the initial glass.

Each preform was stretched on a drawing tower specifically designed for low T_g glasses. Several meters of fibers with approximately a 400 μm diameter were obtained and optically characterized using the FTIR and a MCT (alloy of mercury, cadmium and tellurium) detector. Attenuation measurements were performed in order to compare the optical transmission of the initial glass fibers and the "printed" fibers. The results are shown in Figure 6. A significant increase in optical losses is observed for the "printed" fibers. Thus, when the minimum attenuation of the initial unpurified glass is less than 10 dB/m, and less than 1 dB/m for the purified glass, the minimum attenuation of the unpurified "printed" fiber is around 28 dB/m and around 18 dB/m for the purified "printed" fiber.

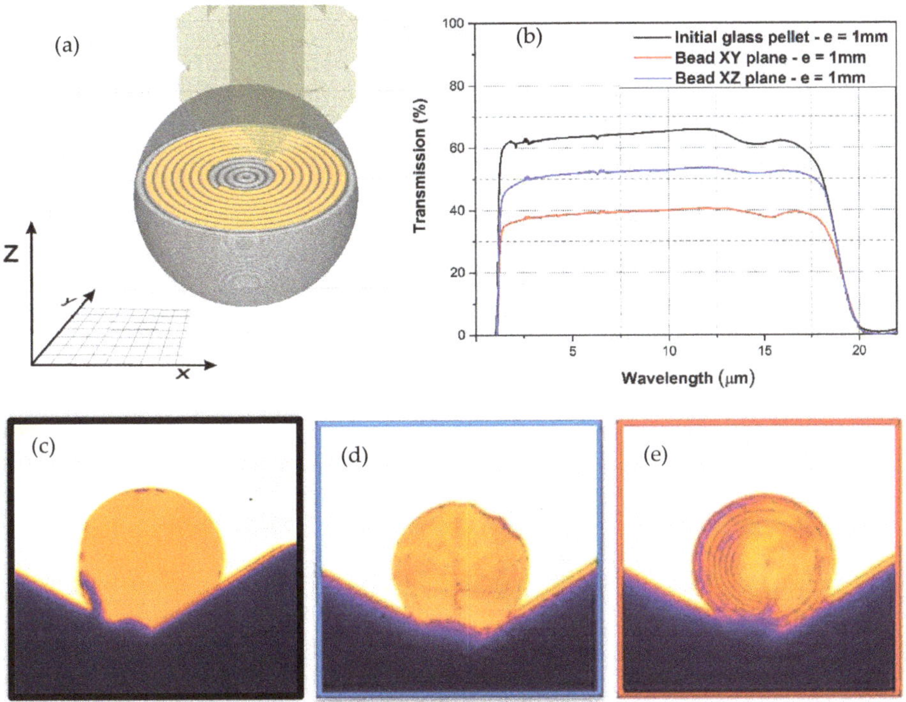

Figure 5. Optical transmission of printed beads: (**a**) numerical view of the 3D printing process, (**b**) initial glass transmission compared to the transmission of 2 disks cut in the XY plan (red curve) and XZ plan (blue curve), respectively. IR camera views in the 7–13 μm window (**c**) initial TAS glass, (**d**) XZ plan, (**e**) XY plan.

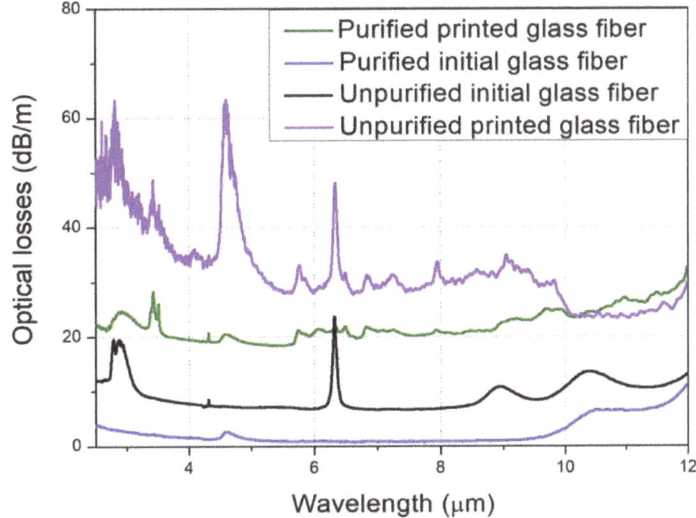

Figure 6. Attenuation spectra of the initial glasses (unpurified and purified glasses, red curve and blue curve respectively) and the "printed" glass fibers (from an unpurified glass and from a purified glass, purple and green respectively)

3.3. Optical Components and Fibers Printing

The printed objects used for the previous characterizations were bulk in nature with a concentric shape. It is therefore interesting to evaluate how the chalcogenide glass behaves when the printed objects are thinner and mainly made up of a straight line, for potentially making smaller components, such as sensors [33,34] by using this innovative 3D printing process.

First, a slot shape object has been printed, as shown in Figure 7a. The lines are around 400 μm wide and 700 μm high. The width corresponds to the diameter of the nozzle (400 μm) and the height is the sum of seven printed layers (7 × 100 μm). The longest lines are equal to 3 cm and the shortest to 1 cm. The second object, as shown in Figure 7b, is a glass comb whose height is around 1 mm, the width of the teeth being close to 800 μm, which corresponds to two printed lines. The length of the teeth is 3 cm.

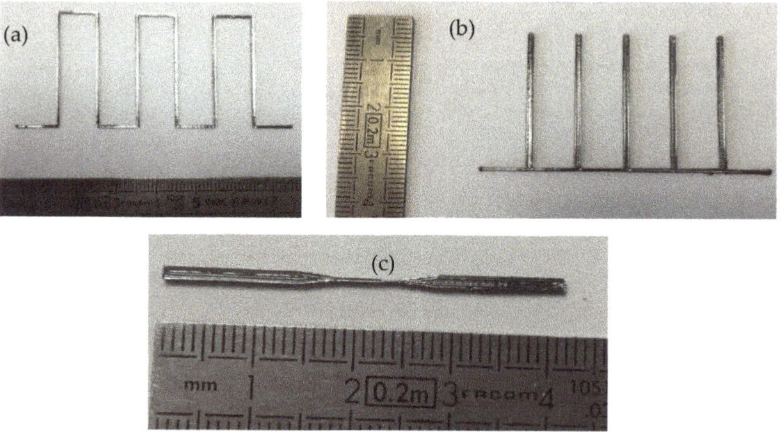

Figure 7. Linear 3D printed chalcogenide glass: (**a**) 700 × 500 μm slots shapes, (**b**) 1000 × 800 μm comb and (**c**) 1000 × 250 μm rectangular taper.

A third object was printed with a 0.25 mm diameter nozzle using the same thermal conditions and feeding rods. It is a straight object with two wide edges of 1000 × 1000 μm^2 over a length of 15 mm connected by a taper of 1000 × 250 μm^2 over a length of 10 mm, as shown in Figure 7c. For these small printed components, variations of about 10% in height and width were measured. These results show that various shapes of chalcogenide glass objects can be obtained by additive manufacturing and demonstrate the strong potential of such a fabrication approach.

Finally, additive manufacturing was applied to the printing of fiber preforms as previously shown in reference [22], with the printing of a hollow-core preform with a clad and six capillaries. After this first proof of concept, more complex designs were printed as shown in Figure 8a. The first preform is a hollow-core preform constituted by 8 capillaries of 1.2-mm diameter. The preform had an outer diameter of 18 mm and a height of 15 mm. This preform was too short to be drawn. A second preform, illustrated in Figure 8b, has been drawn into fiber (Figure 8c). This printed preform was designed with eight half-capillaries as illustrated in Figure 8b. The printing of a such geometry permits one to reduce the diameter of the preform and consequently permits one to increase the printed preform length by using the same volume of glass. The diameter of the semi-circles is 1.8 mm. The overall diameter and height of the preform are 15.2 mm and 30 mm, respectively. This design was realized also for showing the interest of 3D printing that can permit one to obtain geometries that cannot be obtained by the stack and draw technique. However, no light propagation in mid-IR was observed in the hollow core of the fiber drawn from this preform. This is attributed to the final geometry of the as-prepared fiber (Figure 8c) that needs to be optimized to allow mid-IR light transmission (core size, half-capillaries thickness . . .).

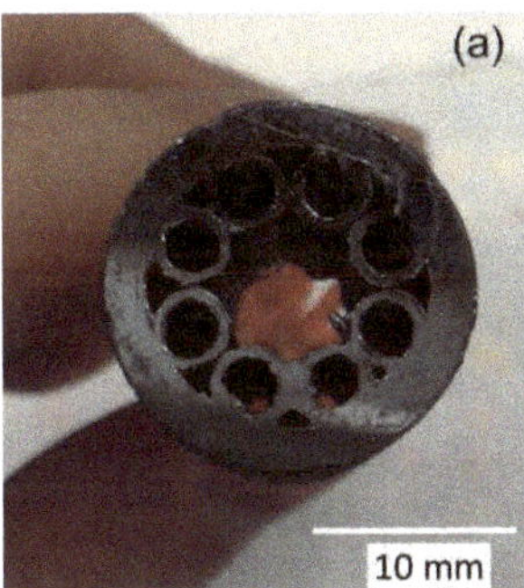

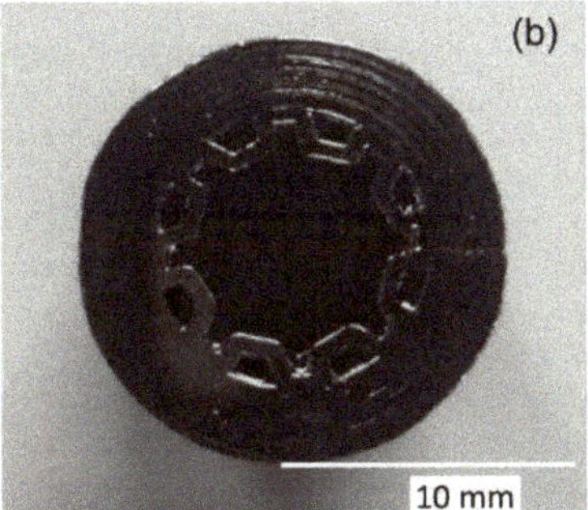

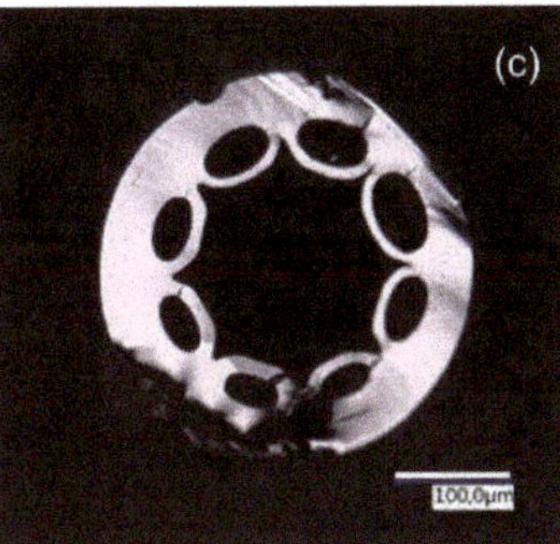

Figure 8. Printed preforms: (**a**) cross-section view of the chalcogenide printed 8 capillaries preform and (**b**) cross-section view of the chalcogenide printed 8 half-capillaries preform (**c**) cross section of a hollow core fiber drawn from the preform shown in (**b**).

4. Discussion

The aim of this work was, firstly, to study the impact of 3D printing by filamentation on the properties of chalcogenide glasses, and secondly, to apply additive manufacturing to produce optical objects and components for mid-infrared applications.

The comparison of the physical and thermal properties between the initial glass and the printed glass, as presented in Table 1, shows no major differences. However, the investigation of the optical properties showed a significant difference between the two processes. Indeed, a significant decrease in transmission was observed both on bulk printed glasses and printed fibers. The transmission of TAS glass obtained by the usual melt–quench process exceeds 60% in the 2–18 µm window, whereas the transmission along the Z axis, which corresponds to the axis of stacking of the layers, is below 40%, and the transmission in a direction perpendicular to the Z axis is 50%. As illustrated in Figure 5d,e, the images recorded with a thermal camera in both directions show the presence of numerous inhomogeneities and scattering defects which are responsible of the decrease in transmission. These inhomogeneities could be refractive index variations or additional absorptions at the interfaces between the printed lines and the printed layers, and could also be the results of small bubbles trapped within the glass, as already observed in previous work [22]. The presence of small crystals in the printed pieces cannot be totally excluded, even though XRD investigations have shown no sign of crystallization due to the process (Figure 4). The infrared images also show that the interfaces between the lines (along the Z axis, Figure 5e) present more defects than the interfaces between the layers (Figure 5d). Indeed, the IR image, presented in Figure 5e, clearly shows concentric lines that are related to the movement of the extrusion head for one printing layer (Figure 5a). In order to improve the transmission of the printed chalcogenide glasses, this effect should be considered in further studies, by changing the 3D printing parameters such as the nozzle size, the layer thickness, the extrusion temperature, the extrusion, and printing rates, which have not yet been investigated.

The study of the optical properties of fibers has also revealed significant differences between the initial glass and the printed glass and in particular shows the impact of the quality of the starting glass. The minimum of attenuation of the initial glass were measured at around 8 dB/m and 1 dB/m for the unpurified glass fiber and purified glass fiber, respectively. The minimum of attenuation of the "printed" fiber made from the unpurified glass is close to 28 dB/m, whereas the preform realized with a purified glass permit one to obtain propagation losses in the fiber lower than 18 dB/m. So, the quality of the printed glass is clearly better when a purified glass is used for printing. However, it can be noticed that the additional losses induced by the printing process are similar. Indeed, the additional losses induced by the 3D printing is 20 dB/m for the unpurified glass and 17 dB/m for the purified glass, which indicates that the majority of the losses are induced by the 3D printing process itself. In addition to the background losses, numerous other additional absorption peaks are observed in the attenuation spectra of both printed glasses drawn into fibers: OH groups and molecular H_2O signatures at 2.9 μm and 6.3 μm respectively, C–H chemical IR signatures at 3.3 μm, other undefined organic compounds signatures at 5.8 μm, 6.8 μm and 7.3 μm, and finally oxidation signatures at 7.9 at 9.1 μm. It is important to note here that the 3D printing process is carried out in air. In these experimental conditions, the origin of the additional absorption bands is probably due to the presence of moisture, oxygen and possible traces of pollution by organic compounds in the printer's enclosure. For the fiber drawn with an unpurified printed preform, the strong increase in the Se–H bond IR signature at 4.55 μm is surprising in comparison to the purified glass fiber spectrum for which the characteristic band of Se–H does not evolve. However, Se–H bonds are the result of a chemical reaction between chalcogenide glass and hydrogen coming mainly from O–H bonds and/or molecular water. In our case, before printing, the unpurified initial glass shows high O–H and molecular water contents, at 2.9 μm and 6.3 μm respectively, whereas the purified glass does not contain any water pollution. The most probable hypothesis is that the water contained in the initial unpurified glass has reacted with the selenium of the glass to form Se–H bonds during printing.

The main interests of additive manufacturing processes are the realization of complex objects that cannot be obtained by other methods but also for mass production and for mass customization with a significant cost reduction. In order to evaluate the potential of our 3D printed method, various shapes and small objects were achieved.

Concerning the objects showing centimetric sizes, it has been shown that several shapes such as disks, cylinders, and beads can be 3D printed with the selected chalcogenide glass. However, the investigations have shown that improvements must be implemented in order to reach the same optical transmission in a printed glass than in a glass synthesized by the usual melt–quenching process. For future studies, the priority improvement should be the use of an inert and dry atmosphere in the printed enclosure to prevent any defects resulting from chemical pollution.

In the present experimental set-up, the resolution of one line (approximately equal to the size of nozzle) is 250 μm or 400 μm and the resolution along the Z axis is 100 μm, which corresponds to the thickness of one layer. For preliminary results, these resolutions seem to be sufficient for printing components with size smaller than 1 mm. These preliminary results also show that it will be possible to exploit the properties of chalcogenide glasses to make infrared sensors. In fact, the printed taper in Figure 7c exhibits a design very close to that of infrared fiber sensors used commercially by DIAFIR company to carry out infrared spectroscopy and medical diagnosis [33,34].

5. Conclusions

The investigations on the thermal properties between the initial glass and the printed glass have shown no major difference. Indeed, the composition of the glass and the glass transition temperature are the same before and after the printing operation. In addition, no sign of a crystallization is reported in the printed glasses. The study has also shown that improvements have to be achieved in order to increase the optical quality of the printed glasses. For example, the use of an inert and dry atmosphere during the printing has been pointed out and this should constitute the main future development axis.

Finally, this work clearly demonstrates that 3D printing could be an innovative approach for elaborating microstructured fibers. Especially, it allows one to obtain new geometries that cannot be realized by any other methods such as stack and draw [35,36], extrusion [37] or molding [38].

This innovative 3D printing method opens the way to many applications involving chalcogenide fiber manufacturing, but also many other optical devices based on chalcogenide glasses, such as chalcogenide sensors for spectroscopy and medical diagnosis.

Author Contributions: Conceptualization, J.T., L.C., R.L., F.C. and G.R.; methodology, A.G., F.C., L.B. and J.C.; formal analysis, E.G. and J.C.; investigation, J.C., R.C. and L.B.; writing—original draft preparation, J.C.; writing—review and editing, J.T., F.C., D.L.C., G.R. and J.L.A.; supervision and project administration, J.T.; funding acquisition, J.T., G.R. and L.B. All authors have read and agreed to the published version of the manuscript.

Funding: This work was funded in part by the European Union through the European Regional Development Fund (ERDF), the Ministry of Higher Education and Research, the French region of Brittany, Rennes Métropole, and the French Délégation Générale pour l'armement (DGA) (grant ANR ASTRID DGA FOM-IR-2-20).

Institutional Review Board Statement: Not applicable.

Informed Consent Statement: Not applicable.

Data Availability Statement: The datasets generated during and/or analysed during the current study are available from the corresponding author on reasonable request.

Acknowledgments: See funding section.

Conflicts of Interest: The authors declare no conflict of interest.

References

1. Zakery, A.; Elliott, S.R. Optical properties and applications of chalcogenide glasses: A review. *J. Non-Cryst. Solids* **2003**, *330*, 1–12. [CrossRef]
2. Adam, J.-L.; Calvez, L.; Troles, J.; Nazabal, V. Chalcogenide Glasses for Infrared Photonics. *Int. J. Appl. Glass Sci.* **2015**, *6*, 287–294. [CrossRef]
3. Heo, J.; Rodrigues, M.; Saggese, S.J.; Sigel, G.H. Remote fiber-optic chemical sensing using evanescent-wave interactions in chalcogenide glass fibers. *Appl. Opt.* **1991**, *30*, 3944–3951. [CrossRef]
4. Adam, J.-L.; Zhang, X. *Chalcogenide Glasses: Preparation, Properties and Applications*; Woodhead Publishing: Cambridge, UK, 2014.
5. Lavanant, E.; Calvez, L.; Cheviré, F.; Rozé, M.; Hingant, T.; Proux, R.; Guimond, Y.; Zhang, X.-H. Radial gradient refractive index (GRIN) infrared lens based on spatially resolved crystallization of chalcogenide glass. *Opt. Mater. Express* **2020**, *10*, 860–867. [CrossRef]
6. Calvez, L. Chalcogenide glasses and glass-ceramics: Transparent materials in the infrared for dual applications. *Comptes Rendus Phys.* **2017**, *18*, 314–322. [CrossRef]
7. Aggarwal, I.D.; Sanghera, J.S. Development and applications of chalcogenide glass optical fibers at NRL. *J. Optoelectron. Adv. Mater.* **2002**, *4*, 665–678.
8. Lucas, J.; Troles, J.; Zhang, X.H.; Boussard-Pledel, C.; Poulain, M.; Bureau, B. Glasses to see beyond visible. *Comptes Rendus Chimie* **2018**, *21*, 916–922. [CrossRef]
9. Toupin, P.; Brilland, L.; Renversez, G.; Troles, J. All-solid all-chalcogenide microstructured optical fiber. *Opt. Express* **2013**, *21*, 14643–14648. [CrossRef] [PubMed]
10. Gibson, I.; Rosen, D.W.; Stucker, B. *Additive Manufacturing Technologies*; Springer: Boston, MA, USA, 2014. [CrossRef]
11. Bandyopadhyay, A.; Bose, S. *Additive Manufacturing*; CRC Press: Boca Raton, FL, USA, 2019.
12. Kumar, S.; Czekanski, A. Roadmap to sustainable plastic additive manufacturing. *Mater. Today Commun.* **2018**, *15*, 109–113. [CrossRef]
13. Ngo, T.D.; Kashani, A.; Imbalzano, G.; Nguyen, K.T.Q.; Hui, D. Additive manufacturing (3D printing): A review of materials, methods, applications and challenges. *Compos. Part B Eng.* **2018**, *143*, 172–196. [CrossRef]
14. Murr, L.E.; Gaytan, S.M.; Ramirez, D.A.; Martinez, E.; Hernandez, J.; Amato, K.N.; Shindo, P.W.; Medina, F.R.; Wicker, R.B. Metal Fabrication by Additive Manufacturing Using Laser and Electron Beam Melting Technologies. *J. Mater. Sci. Technol.* **2012**, *28*, 1–14. [CrossRef]
15. Zocca, A.; Colombo, P.; Gomes, C.M.; Günster, J. Additive Manufacturing of Ceramics: Issues, Potentialities, and Opportunities. *J. Am. Ceram. Soc.* **2015**, *98*, 1983–2001. [CrossRef]
16. Marchelli, G.; Prabhakar, R.; Storti, D.; Ganter, M. The guide to glass 3D printing: Developments, methods, diagnostics and results. *Rapid Prototyp. J.* **2011**, *17*, 187–194. [CrossRef]
17. Kotz, F.; Arnold, K.; Bauer, W.; Schild, D.; Keller, N.; Sachsenheimer, K.; Nargang, T.M.; Richter, C.; Helmer, D.; Rapp, B.E. Three-dimensional printing of transparent fused silica glass. *Nature* **2017**, *544*, 337–339. [CrossRef]
18. Chu, Y.; Fu, X.; Luo, Y.; Canning, J.; Tian, Y.; Cook, K.; Zhang, J.; Peng, G.-D. Silica optical fiber drawn from 3D printed preforms. *Opt. Lett.* **2019**, *44*, 5358–5361. [CrossRef]

19. Turner, B.N.; Strong, R.; Gold, S.A. A review of melt extrusion additive manufacturing processes: I. Process design and modeling. *Rapid Prototyp. J.* **2014**, *20*, 192. [CrossRef]
20. Turner, B.N.; Gold, S.A. A review of melt extrusion additive manufacturing processes: II. Materials, dimensional accuracy, and surface roughness. *Rapid Prototyp. J.* **2015**, *21*, 250. [CrossRef]
21. Baudet, E.; Ledemi, Y.; Larochelle, P.; Morency, S.; Messaddeq, Y. 3D-printing of arsenic sulfide chalcogenide glasses. *Opt. Mater. Express* **2019**, *9*, 2307–2317. [CrossRef]
22. Carcreff, J.; Chevïré, F.; Galdo, E.; Lebullenger, R.; Gautier, A.; Adam, J.L.; Coq, D.L.; Brilland, L.; Chahal, R.; Renversez, G.; et al. Mid-infrared hollow core fiber drawn from a 3D printed chalcogenide glass preform. *Opt. Mater. Express* **2021**, *11*, 198–209. [CrossRef]
23. Zaki, R.M.; Strutynski, C.; Kaser, S.; Bernard, D.; Hauss, G.; Faessel, M.; Sabatier, J.; Canioni, L.; Messaddeq, Y.; Danto, S.; et al. Direct 3D-printing of phosphate glass by fused deposition modeling. *Mater. Des.* **2020**, *194*, 108957. [CrossRef]
24. Available online: https://marlinfw.org (accessed on 6 February 2021).
25. Shiryaev, V.S.; Adam, J.L.; Zhang, X.H.; Boussard-Pledel, C.; Lucas, J.; Churbanov, M.F. Infrared fibers based on Te-As-Se glass system with low optical losses. *J. Non-Cryst. Solids* **2004**, *336*, 113–119. [CrossRef]
26. Fulcher, G.S. Analysis of recent measurements of the viscosity of glasses. *J. Am. Ceram. Soc.* **1925**, *8*, 339–355. [CrossRef]
27. Yoon, Y.I.; Park, K.; Lee, S.J.; Park, W.H. Fabrication of Microfibrous and Nano-/Microfibrous Scaffolds: Melt and Hybrid Electrospinning and Surface Modification of Poly(L-lactic acid) with Plasticizer. *BioMed. Res. Int.* **2013**, *2013*, 309048. [CrossRef]
28. Gunduz, I.E.; McClain, M.S.; Cattani, P.; Chiu, G.T.C.; Rhoads, J.F.; Son, S.F. 3D printing of extremely viscous materials using ultrasonic vibrations. *Addit. Manuf.* **2018**, *22*, 98–103. [CrossRef]
29. Cheng, C.-C. Refractive index measurement by prism autocollimation. *Am. J. Phys.* **2014**, *82*, 214–216. [CrossRef]
30. Caillaud, C.; Renversez, G.; Brilland, L.; Mechin, D.; Calvez, L.; Adam, J.-L.; Troles, J. Photonic Bandgap Propagation in All-Solid Chalcogenide Microstructured Optical Fibers. *Materials* **2014**, *7*, 6120–6129. [CrossRef]
31. Conseil, C.; Coulombier, Q.; Boussard-Pledel, C.; Troles, J.; Brilland, L.; Renversez, G.; Mechin, D.; Bureau, B.; Adam, J.L.; Lucas, J. Chalcogenide step index and microstructured single mode fibers. *J. Non-Cryst. Solids* **2011**, *357*, 2480–2483. [CrossRef]
32. Meneghetti, M.; Caillaud, C.; Chahal, R.; Galdo, E.; Brilland, L.; Adam, J.-L.; Troles, J. Purification of Ge-As-Se ternary glasses for the development of high quality microstructured optical fibers. *J. Non-Cryst. Solids* **2019**, *503*, 84–88. [CrossRef]
33. Bensaid, S.; Kachenoura, A.; Costet, N.; Bensalah, K.; Tariel, H.; Senhadji, L. Noninvasive detection of bladder cancer using mid-infrared spectra classification. *Expert Syst. Appl.* **2017**, *89*, 333–342. [CrossRef]
34. Le Corvec, M.; Jezequel, C.; Monbet, V.; Fatih, N.; Charpentier, F.; Tariel, H.; Boussard-Pledel, C.; Bureau, B.; Loreal, O.; Sire, O.; et al. Mid-infrared spectroscopy of serum, a promising non-invasive method to assess prognosis in patients with ascites and cirrhosis. *PLoS ONE* **2017**, *12*, 15. [CrossRef]
35. Russell, P. Photonic Crystal Fibers. *Science* **2003**, *299*, 358. [CrossRef] [PubMed]
36. Brilland, L.; Smektala, F.; Renversez, G.; Chartier, T.; Troles, J.; Nguyen, T.N.; Traynor, N.; Monteville, A. Fabrication of complex structures of Holey Fibers in chalcogenide glass. *Opt. Express* **2006**, *14*, 1280–1285. [CrossRef] [PubMed]
37. Gattass, R.R.; Rhonehouse, D.; Gibson, D.; McClain, C.C.; Thapa, R.; Nguyen, V.Q.; Bayya, S.S.; Weiblen, R.J.; Menyuk, C.R.; Shaw, L.B.; et al. Infrared glass-based negative-curvature anti-resonant fibers fabricated through extrusion. *Opt. Express* **2016**, *24*, 25697–25703. [CrossRef] [PubMed]
38. Coulombier, Q.; Brilland, L.; Houizot, P.; Chartier, T.; Nguyen, T.N.; Smektala, F.; Renversez, G.; Monteville, A.; Méchin, D.; Pain, T.; et al. Casting method for producing low-loss chalcogenide microstructured optical fibers. *Opt. Express* **2010**, *18*, 9107–9112. [CrossRef] [PubMed]

Article

Photoionization-Induced Broadband Dispersive Wave Generated in an Ar-Filled Hollow-Core Photonic Crystal Fiber

Jianhua Fu [1,2,†], Yifei Chen [1,2,†], Zhiyuan Huang [1,2,*], Fei Yu [3,4], Dakun Wu [4], Jinyu Pan [1,2], Cheng Zhang [4], Ding Wang [1], Meng Pang [1,4] and Yuxin Leng [1,4,*]

1 State Key Laboratory of High Field Laser Physics and CAS Center for Excellence in Ultra-Intense Laser Science, Shanghai Institute of Optics and Fine Mechanics (SIOM), Chinese Academy of Sciences (CAS), Shanghai 201800, China; fujianhua@siom.ac.cn (J.F.); yfchen@siom.ac.cn (Y.C.); panjinyu@siom.ac.cn (J.P.); wangding@siom.ac.cn (D.W.); pangmeng@siom.ac.cn (M.P.)
2 Center of Materials Science and Optoelectronics Engineering, University of Chinese Academy of Sciences, Beijing 100049, China
3 R&D Center of High Power Laser Components, Shanghai Institute of Optics and Fine Mechanics, Chinese Academy of Sciences, Shanghai 201800, China; yufei@siom.ac.cn
4 Hangzhou Institute for Advanced Study, Chinese Academy of Sciences, Hangzhou 310024, China; wudakun@ucas.ac.cn (D.W.); zhangcheng@siom.ac.cn (C.Z.)
* Correspondence: huangzhiyuan@siom.ac.cn (Z.H.); lengyuxin@mail.siom.ac.cn (Y.L.)
† These authors contributed equally to this work.

Abstract: The resonance band in hollow-core photonic crystal fiber (HC-PCF), while leading to high-loss region in the fiber transmission spectrum, has been successfully used for generating phase-matched dispersive wave (DW). Here, we report that the spectral width of the resonance-induced DW can be largely broadened due to plasma-driven blueshifting soliton. In the experiment, we observed that in a short length of Ar-filled single-ring HC-PCF the soliton self-compression and photoionization effects caused a strong spectral blueshift of the pump pulse, changing the phase-matching condition of the DW emission process. Therefore, broadening of DW spectrum to the longer-wavelength side was obtained with several spectral peaks, which correspond to the generation of DW at different positions along the fiber. In particular, we numerically used the super-Gauss windows with different central wavelengths to filter out these DW spectral peaks and studied the time-domain characteristics of these peaks respectively using Fourier transform method. We observed that these multiple-peaks on the DW spectrum have different delays in the time domain, which is in good agreement with our theoretical prediction. More interestingly, we found that the broadband DW with several spectral peaks can be compressed to ~29 fs after proper dispersion compensation. The results reported here, on the one hand, provide some useful insights into the resonance-induced DW generation process in gas-filled HC-PCFs. On the other hand, the DW-emission mechanism could be used to generate the ultrashort light sources with a wide spectral range through using the proper design of the resonance bands of the HC-PCFs, which has many applications in the ultrafast related experiments.

Keywords: hollow-core photonic crystal fiber; soliton; photoionization; dispersive wave

Citation: Fu, J.; Chen, Y.; Huang, Z.; Yu, F.; Wu, D.; Pan, J.; Zhang, C.; Wang, D.; Pang, M.; Leng, Y. Photoionization-Induced Broadband Dispersive Wave Generated in an Ar-Filled Hollow-Core Photonic Crystal Fiber. *Crystals* **2021**, *11*, 180. https://doi.org/10.3390/cryst11020180

Academic Editors: Nicolas Y. Joly and David Novoa

Received: 29 December 2020
Accepted: 5 February 2021
Published: 12 February 2021

1. Introduction

The low-loss broadband-guiding gas-filled hollow-core photonic crystal fibers (HC-PCFs), as ideal nonlinear platforms, are widely used for studying the soliton-related dynamics such as soliton self-compression, soliton-plasma interactions and phase-matched dispersive wave (DW) [1,2]. In particular, the phase-matched DW generated in gas-filled HC-PCFs, arising from a nonlinear energy transfer from a self-compressed soliton [3–6], has attracted great research interests in the past ten years [7]. The phase-matched DW can not only be generated in a wide wavelength range such as ultraviolet (UV) [8] and mid-infrared (MIR) spectral regions [9], but its wavelength can also be effectively tuned, which makes this kind of light sources have many applications, especially in ultrafast spectroscopy [10,11].

In the gas-filled HC-PCFs, the combined effects of self-compressed soliton and high-order dispersion can result in phase-matched DW generated in the UV region [1,7]. As early as ten years ago, Joly et al. demonstrated the efficient emission of DW in the deep-UV region by using an Ar-filled kagomé-style HC-PCF [8]. The generated bright spatially coherent deep-UV laser source is tunable from 200 to 320 nm through varying the pulse energy and gas pressure. Belli et al. used a short length of hydrogen-filled kagomé-style HC-PCF to develop the generation of vacuum-UV (VUV) to near-infrared (NIR) supercontinuum [12]. The experimental result showed that a strong VUV DW emission generated at 182 nm on the trailing edge of the pulse, and it also proved that kagomé-style HC-PCF works well in the VUV spectral region. Simultaneously, Ermolov et al. reported on the generation of a three-octave-wide supercontinuum from VUV to NIR region in a He-filled kagomé-style HC-PCF [13]. The VUV pulses generated through DW emission is tunable from 120 to 200 nm, with efficiencies >1% and VUV pulse energy >50 nJ. It should be pointed out that in the strong-field regime, the plasma caused by gas ionization modifies the fiber dispersion, allowing phase-matched DW generation in the MIR region. Novoa et al. first predicted the generation of MIR DW and established a new model to explain it well [14]. Köttig et al. first experimentally demonstrated the existence of MIR DW and realized a 4.7-octave-wide supercontinuum from 180 nm to 4.7 μm, with up to 1.7 W of total average power [9].

The phase-matched DW can not only be generated in the UV and MIR spectral regions, but also in the visible and NIR regions. Sollapur et al. experimentally reported on the generation of a three-octave-wide supercontinuum from 200 nm to 1.7 μm at an output energy of ~23 μJ in a Kr-filled HC-PCF [15]. Simulations showed that the spectra generated in the visible and NIR resonance bands are closely related to the emission of the phase-matched DW. Tani et al. further proved that the narrow-band spectral peaks generated in the resonance bands are based on the phase-matched DW emission due to the anti-crossing dispersion [16]. Recently, Meng et al. demonstrated the generation of NIR DW in the resonance bands of an Ar-filled kagomé-style HC-PCF [17]. In our recent experiments [18,19], we reported the high-efficiency emission of phase-matched DW in the visible spectral region using a He-filled single-ring (SR) HC-PCF. As the input pulse energy increases, the central wavelength of the plasma-driven blueshifting soliton (BS) [20,21] is close to the resonance band of the fiber, high-efficiency energy transfer from the pump light to the DW can be triggered.

In this work, we experimentally demonstrated the photoionization-induced broad-band DW in a 25-cm-long Ar-filled SR HC-PCF. In a certain pulse energy region (from 1 μJ to 1.5 μJ), we observed that soliton self-compression of the input pulse results in a spectral expansion that overlaps with resonant DW frequencies, leading to a narrow-band DW spectral peak in the first resonance band of the SR HC-PCF. At high pulse energy levels (from 1.5 μJ to 3.2 μJ), we observed that the plasma-driven BS can further excite multiple DW peaks, which leads to the broadening of DW spectrum to the longer-wavelength side. In addition, we theoretically investigated the time-domain characteristics of these spectral peaks filtered by the super-Gauss windows using Fourier transform method, and numerically showed that via suitable dispersion compensation the DW pulses can be compressed to ~29 fs.

2. Experimental Set-Up

The experimental set-up is shown in Figure 1. The 800 nm, ~45 fs, 0.3 mJ pulses from a commercial Ti:Sapphire laser system (Legend Elite, Coherent, CA, USA) were coupled into a 1-m-long hollow-core fiber (HCF) by using a concave mirror with a focal length of 1 m. The HCF has a core diameter of 250 μm and was placed in a gas cell filled with 212 mbar Ar. The spectrum of the pulses after propagating the Ar-filled HCF was broadened due to self-phase modulation (SPM), and the energy transmission was measured to be ~70%. Several pairs of chirped mirrors (CMs) were not only used to compensate the output pulses from the gas-filled HCF, but also to pre-compensate the dispersion introduced by some optical elements, including half-wave plate (HWP), wire grid polarizer (WGP), and even plano-convex lens (PCL) and the window at the input port of the second gas cell. The first pair of CMs (Layertec) provides -40 fs^2 group delay dispersion (GDD) per bounce, and the last three pairs of CMs (Thorlabs) offer -54 fs^2 GDD per bounce. The total GDD provided by these CMs is about -564 fs^2.

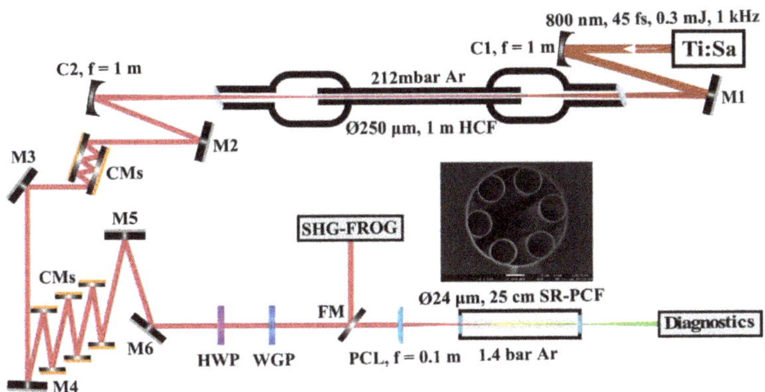

Figure 1. Schematic of the experimental set-up (Visio 2013, Microsoft, WA, USA). M1–M6, silver mirrors; C1-C2, concave mirrors; CMs, chirped mirrors; HWP, half-wave plate; WGP, wire grid polarizer; FM, flipping mirror; PCL, plano-convex lens. The inset in the set-up represents SEM of the SR-PCF with a core diameter of 24 μm and a wall thickness of ~0.26 μm.

In the experiments, we used a home-built second-harmonic-generation frequency-resolved optical gating (SHG-FROG, Shanghai, China) to measure the input pulses in front of the PCL. The retrieved trace with an error of 0.6%, as shown in Figure 2b, agrees well with the measured trace in Figure 2a. The retrieved temporal and spectral intensities of the input pulses and their phase are plotted in Figure 2c,d. The measured input pulse duration is ~18.7 fs [full width at half-maximum (FWHM)]. Then the compressed pulses were launched into a 25-cm-long SR HC-PCF filled with 1.4 bar Ar using a coated PCL with a focal length of 0.1 m. The insert in Figure 1 indicates the scanning electron micrograph (SEM) of the SR HC-PCF that has a core diameter of 24 μm and a wall thickness of ~0.26 μm. In Figure 2e, we plot the simulated (green solid line) and measured (purple solid line) fiber losses of the fundamental optical mode HE_{11} of the SR HC-PCF. The simulated fiber loss was calculated by the bouncing ray (BR) model [22]. Figure 2f shows the dispersion curve of Ar-filled SR HC-PCF at gas pressure of 1.4 bar calculated by the fully analytical model created by Zeisberger and Schmidt (ZS) [23]. In both Figure 2e,f, the cyan bars point out the first resonance band of the SR HC-PCF.

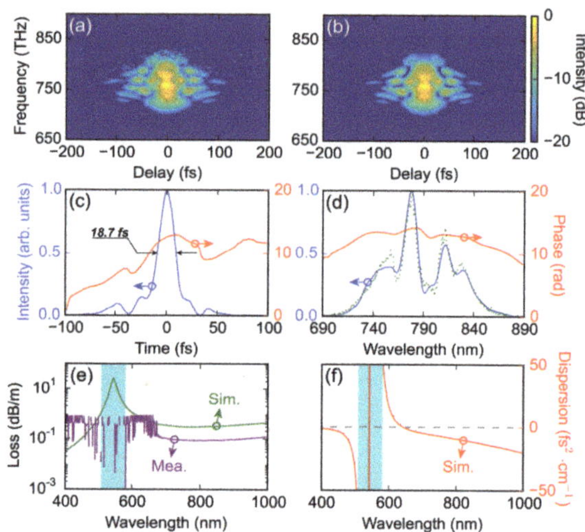

Figure 2. Plotted using the open source software Python(x,y). (**a**,**b**) Measured and retrieved SHG-FROG traces of the pulses in front of the focusing lens. (**c**,**d**) Retrieved temporal and spectral profiles (blue solid lines) and the corresponding phase (red solid lines). The green dotted line in panel (**d**) represents the measured reference spectrum. (**e**) Simulated (green solid line) and measured (purple solid line) fiber losses of the fundamental mode HE_{11} of the SR HC-PCF. (**f**) Simulated dispersion (red solid line) of the SR HC-PCF filled with 1.4-bar Ar gas. The simulated fiber loss and dispersion are calculated through the BR and ZS models, respectively. In both (**e**,**f**), the cyan bars show the first resonant spectral region of the SR HC-PCF.

3. Experimental Results and Analysis

As show in Figure 3a, we measured the spectral evolutions after a 25-cm-long Ar-filled SR HC-PCF (24 μm core diameter and ~0.26 μm wall thickness) at gas pressure of 1.4 bar as a function of input pulse energy from 0.5 μJ to 3.2 μJ. Moreover, in order to better understand the mechanism of photoionization-induced broadband DW generated in the Ar-filled SR HC-PCF, we numerically simulated the propagation of ultrashort pulses along the SR HC-PCF using the single-mode unidirectional pulse propagation equation (UPPE) [24–26], the corresponding results are shown in Figure 3b. In the simulations, we used the pulses measured by SHG-FROG as the input. The BR and ZS models were used to simulate the fiber loss and dispersion, respectively. The gas ionization was also included in the UPPE, calculated by the Perelomov-Popov-Terent'ev model [27]. The overall agreement between numerical results and experimental results is qualitatively good. According to the magnitude of incident pulse energy, we investigate the influence of the soliton compression (SC) and plasma-driven BS on resonance-induced DW from four stages (marked as the white Roman numbers i, ii, iii and iv), which are separated by white dashed lines.

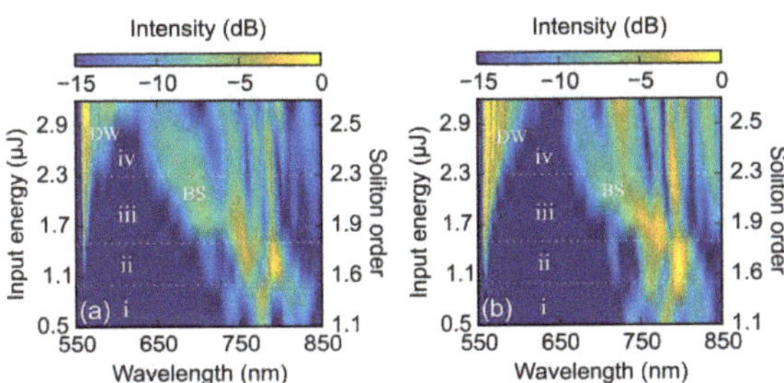

Figure 3. Plotted using the open source software Python(x,y). Measured (**a**) and simulated (**b**) spectral evolutions after propagating a 25-cm-long SR HC-PCF with 24-μm core diameter and 1.4-bar Ar as a function of input pulse energy. In both (**a**,**b**), DW = dispersive wave, BS = blueshifting soliton.

At low pulse energy levels (from 0.5 μJ to 1 μJ), the pulses undergo the SC process due to the combined effects of SPM and waveguide-induced anomalous dispersion. As shown in Figure 4a,b, we plot the simulated temporal and spectral evolutions in a 25-cm-long Ar-filled SR HC-PCF at input energy of 0.5 μJ. As the propagation distance increases, the peak power of the self-compressed pulses increases rapidly, resulting in a rapid accumulation of plasma [marked as the white circle lines in Figure 4b] due to gas ionization. In Figure 4c, we can see that the simulated spectrum (red solid line) at the fiber output is in good agreement with the experimental spectrum (green shaded region). At certain pulse energy levels (from 1 μJ to 1.5 μJ), the SC process leads to a wider pulse spectrum, which results in the frequency to overlap with the linear wave, and then emits the phase-matched DW. At input pulse energy of 1.1 μJ, the self-compressed pulses reach the maximum soliton compression point (MSCP) at the position of ~7.5 cm. Meanwhile, the pulses obtain the maximum spectral broadening and trigger the phase-matched DW generation, as shown in Figure 4d,e. The simulated output spectrum still agrees well with the measured spectrum [see Figure 4f]. At high pulse energy levels (from 1.5 μJ to 2.3 μJ), the plasma-driven BS not only enhances the energy transfer from the pump light to the DW, but also causes the spectral redshift of the DW spectrum, as shown in Figure 4g–i at input energy of 2 μJ. The central wavelength of the BS is close to the resonance band of the SR HC-PCF, which leads to the phase-matched DW in the long-wavelength region. This has been demonstrated in the recent experiments, and a good explanation is given using the phase-matched condition between soliton and linear wave [19]. In Figure 4i, we can observe that the shift of the blueshifting spectrum to the shorter-wavelength region is smaller than the measured spectrum. This may be caused by the spectral recoil effect [28]. As the propagation distance increases, the soliton emitting DW gradually loses energy, transferring it to the DW. To conserve the overall energy of the photons, the central frequency of the soliton is shifted to the opposite direction of the DW radiation. Moreover, in the experiment, since the coupling efficiency drops rapidly due to gas ionization effect at high pulse energy levels [21], the actual pulse energy launched into the Ar-filled SR HC-PCF is less than the value in the simulation, resulting in the spectral recoil effect in the simulation being greater than in the experiment. In addition, this disagreement may also be due to the resonance band of the SR HC-PCF not being exactly the same in the experiment and simulation. At a higher pulse energy region (from 2.3 μJ to 3.2 μJ), the plasma-driven BS can further trigger the broadening of DW spectrum to the longer-wavelength side, which leads to an overlap between the blueshifting spectrum and the DW [see Figure 4j–l with an input energy of 2.6 μJ].

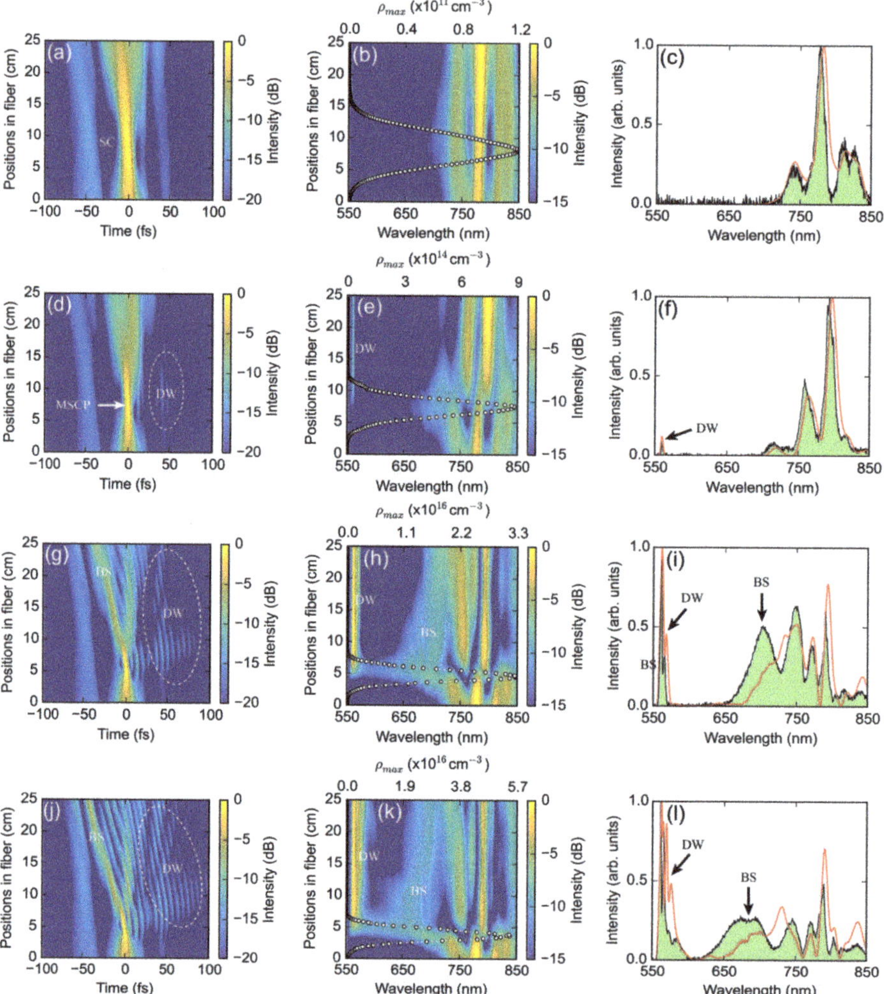

Figure 4. Plotted using the open source software Python(x,y). Simulated temporal (**a**,**d**,**g**,**j**) and spectral (**b**,**e**,**h**,**k**) evolutions in a 25-cm-long Ar-filled SR-PCF with 24-μm core diameter and 1.4-bar gas pressure at different input energies. (**a**–**l**) correspond to input pulse energies of 0.5 μJ, 1.1 μJ, 2.0 μJ and 2.6 μJ, respectively. The measured (green shaded region) and simulated (red solid line) normalized spectral intensities for different input energies at the output of the SR HC-PCF are shown in (**c**,**f**,**i**,**l**). The white circle lines in panels (**b**,**e**,**h**,**k**) point out the maximum plasma density. In (**a**), SC = soliton compression, in (**d**), MSCP = maximum soliton compression point.

The phase-matched DW is usually generated in the positive direction of the time axis. In Figure 5a, we used a super-Gauss window to filter out the leftmost DW spectral band (marked as red dashed line, called DW-SB1), and the green solid line corresponds to the simulated spectrum at the output of the SR HC-PCF with the input pulse energy of 2.6 μJ. Through the Fourier transform of DW-SB1, the corresponding temporal profile in Figure 5b is located on the right side of the time axis and shows a long pulse duration of ~198 fs at FWHM due to a narrow spectral width. However, the plasma-driven BS located in the anomalous dispersion region has a higher group velocity than input pulses, so that the blueshifting pulses accelerate as the propagation distance increases. This causes the DW radiation to approach the zero point of the time axis, and even appear in the negative direction, as shown in Figure 5c–f. The pulse durations corresponding to the DW-SB2 and DW-SB3 are ~171 fs and ~36 fs, respectively. In Figure 5h, we also plot the temporal intensity of BS [see Figure 5g], and the pulse duration is about 13 fs. Figure 5j shows the complete temporal profile of the output pulses. The main peak is the BS, and its peak position is about 44 fs, consistent with Figure 5h. In addition, the left side of the main peak is the existing pedestal of the incident pulses [see Figure 2c], while the oscillations on the right side are the DW radiation and residual pump light. The self-compressed pulse duration can be as short as ~6 fs. These results can also be observed in Figure 4j.

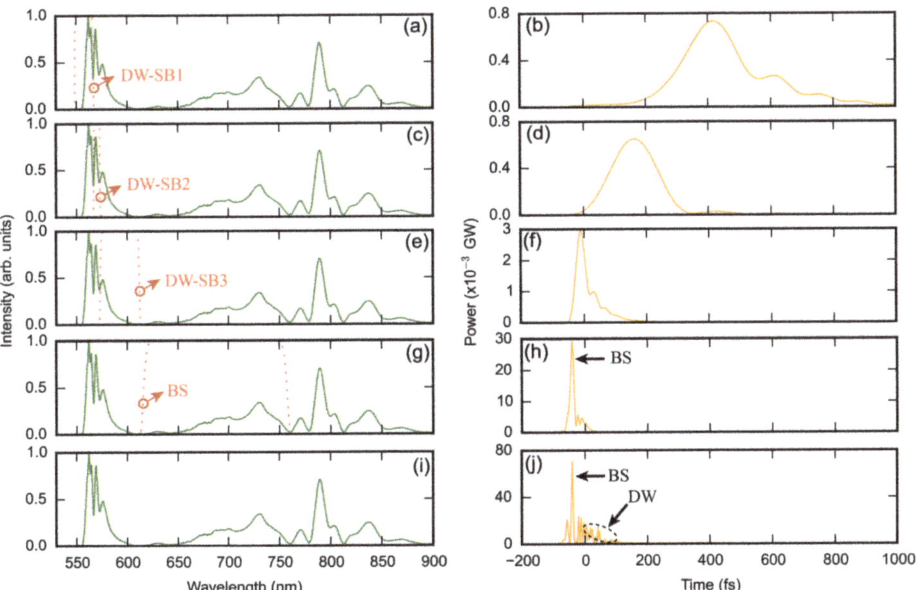

Figure 5. Plotted using the open source software Python(x,y). The green solid lines represent the simulated spectrum output of a 25-cm-long SR HC-PCF at an input energy of 2.6 μJ. In (**b**,**d**,**f**,**h**), the orange solid lines indicate the temporal profiles of the pulses through Fourier transforming the spectral bands (marked as red dashed lines) filtered by super-Gauss windows, and in (**j**), the temporal profile of the pulse obtained by Fourier transforming the entire output spectrum (**i**). In (**a**,**c**,**e**), DW-SB = dispersive wave spectral band and in (**g**), BS = blueshifting soliton.

Figure 6a shows the broadband DW spectrum filtered from the output spectrum of Figure 5i through using a super-Gauss window from 550 nm to 610 nm, and the corresponding temporal profile after the Fourier transform is plotted in Figure 6b. Although the pulse duration at FWHM is ~30 fs, the pulses exhibit long-decay pedestals on the trailing edge due to the DW radiation generated at different positions in the fiber. We found that after −4634 fs^2 GDD and −44,827 fs^3 third-order dispersion (TOD) compensation, these pedestals can be compressed into the main peak, showing a clean pulse leading edge. The compressed pulse duration is ~29 fs, which is very close to the Fourier transform limit (FTL) of ~24 fs. It should be noted that the GDD and TOD compensation amounts in the visible spectral region are both within the compensation range of the device called Dazzler.

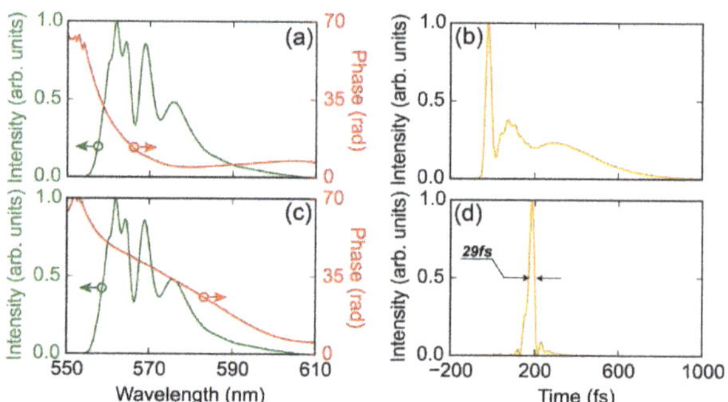

Figure 6. Plotted using the open source software Python(x,y). (**a**) The spectrum (green solid line) filtered from the output spectrum of Figure 5i by using a super-Gauss window from 550 nm to 610 nm, and the corresponding phase is marked as red solid line. After GDD and TOD compensation, the spectrum and phase are shown in panel (**c**). (**b**,**d**) The temporal profiles of the pulses by Fourier transforming the corresponding spectra.

4. Conclusions

In conclusion, we experimentally and numerically demonstrated the photoionization-induced broadband phase-matched DW with several spectral peaks generated in the resonance band of a 25-cm-long Ar-filled SR HC-PCF with core diameter of 24 μm and gas pressure of 1.4 bar. At certain pulse energy levels, the soliton self-compression process can result in a narrow-band DW spectral peak generated near the wavelength of ~550 nm. At high pulse energy levels, the plasma-driven BS gradually moves to the shorter-wavelength region as the propagation distance increases, leading to the broadening of DW spectrum to the longer-wavelength side. Theses experimental observations are well confirmed by the numerical results simulated through well-known UPPE model. Moreover, through using the super-Gauss windows to filter out the DW spectral peaks, we investigated the time-domain characteristics of the DW and better understood the broadband DW generation process. Furthermore, we found that after suitable GDD and TOD compensation, the DW can be compressed together in the time domain, enabling a good pulse quality. The compressed pulse duration is as short as ~29 fs, approaching to the FTL duration of ~24 fs. These experimental results not only offer some useful insights into the resonance-induced DW radiation in gas-filled HC-PCFs, but also provide a simple and effective method for generating the ultrashort light sources in the visible region. Through properly designing the resonance bands of the HC-PCFs, the resonance-induced ultrashort DW light sources could be extended to UV/MIR spectral regions, which has many applications in the ultrafast pump-probe experiments. In the future, we will pay more attention to the temporal and spatial measurements of the broadband DW with multi-peak structure generated in the resonance bands of the HC-PCFs.

Author Contributions: Conceptualization, J.F. and Y.C.; methodology, Y.C. and Z.H.; software, Y.C. and Z.H.; validation, Y.C. and Z.H.; formal analysis, Y.C. and Z.H.; investigation, J.F. and Y.C.; resources, Y.C. and Z.H.; data curation, Z.H. and Y.L.; writing—original draft preparation, J.F., Z.H. and M.P.; writing—review and editing, F.Y., D.W. (Dakun Wu), J.P., C.Z., D.W. (Ding Wang) and Y.L.; visualization, J.F. and Y.C.; supervision, Z.H. and Y.L.; project administration, Y.L.; funding acquisition, Y.L. All authors have read and agreed to the published version of the manuscript.

Funding: Zhangjiang Laboratory Construction and Operation Project, grant number 20DZ2210300; National Key R&D Program of China, grant number 2017YFE0123700; National Natural Science Foundation of China, grant number 61925507; Program of Shanghai Academic/Technology Research Leader, grant number 18XD1404200; Strategic Priority Research Program of the Chinese Academy of Sciences, grant number XDB1603; Major Project Science and Technology Commission of Shanghai Municipality, grant number 2017SHZDZX02.

Institutional Review Board Statement: Not applicable.

Informed Consent Statement: Not applicable.

Data Availability Statement: Not applicable.

Conflicts of Interest: The authors declare no conflict of interest.

References

1. Travers, J.C.; Chang, W.; Nold, J.; Joly, N.Y.; Russell, P.S.J. Ultrafast nonlinear optics in gas-filled hollow-core photonic crystal fibers. *J. Opt. Soc. Am. B* **2011**, *28*, A11–A26. [CrossRef]
2. Russell, P.S.J.; Hölzer, P.; Chang, W.; Abdolvand, A.; Travers, J.C. Hollow-core photonic crystal fibres for gas-based nonlinear optics. *Nat. Photonics* **2014**, *8*, 278–286. [CrossRef]
3. Wai, P.K.A.; Menyuk, C.R.; Lee, Y.C.; Chen, H.H. Nonlinear pulse propagation in the neighborhood of the zero-dispersion wavelength of monomode optical fibers. *Opt. Lett.* **1986**, *11*, 464–466. [CrossRef]
4. Karpman, V.I. Radiation by solitons due to higher-order dispersion. *Phys. Rev. E* **1993**, *47*, 2073–2082. [CrossRef]
5. Akhmediev, N.; Karlsson, M. Cherenkov radiation emitted by solitons in optical fibers. *Phys. Rev. A* **1995**, *51*, 2602–2607. [CrossRef] [PubMed]
6. Erkintalo, M.; Xu, Y.Q.; Murdoch, S.G.; Dudley, J.M.; Genty, G. Cascaded Phase Matching and Nonlinear Symmetry Breaking in Fiber Frequency Combs. *Phys. Rev. Lett.* **2012**, *109*, 223904. [CrossRef]
7. Markos, C.; Travers, J.C.; Abdolvand, A.; Eggleton, B.J.; Bang, O. Hybrid photonic-crystal fiber. *Rev. Mod. Phys.* **2017**, *89*, 045003. [CrossRef]
8. Joly, N.Y.; Nold, J.; Chang, W.; Hölzer, P.; Nazarkin, A.; Wong, G.K.L.; Biancalana, F.; Russell, P.S.J. Bright Spatially Coherent Wavelength Tunable Deep-UV Laser Source Using an Ar-Filled Photonic Crystal Fiber. *Phys. Rev. Lett.* **2011**, *106*, 203901. [CrossRef] [PubMed]
9. Köttig, F.; Novoa, D.; Tani, F.; Günendi, M.C.; Cassataro, M.; Travers, J.C.; Russell, P.S.J. Mid-infrared dispersive wave generation in gas-filled photonic crystal fibre by transient ionization-driven changes in dispersion. *Nat. Commun.* **2017**, *8*, 813. [CrossRef]
10. Zewail, A.H. Femtochemistry: Atomic-Scale Dynamics of the Chemical Bond. *J. Phys. Chem. A* **2000**, *104*, 5660. [CrossRef]
11. Stolow, A.; Bragg, A.E.; Neumark, D.M. Femtosecond Time-Resolved Photoelectron Spectroscopy. *Chem. Rev.* **2004**, *104*, 1719. [CrossRef]
12. Belli, F.; Abdolvand, A.; Chang, W.; Travers, J.C.; Russell, P.S.J. Vacuum-ultraviolet to infrared supercontinuum in hydrogen-filled photonic crystal fibe. *Optica* **2015**, *2*, 292–300. [CrossRef]
13. Ermolov, A.; Mak, K.F.; Frosz, M.H.; Travers, J.C.; Russell, P.S.J. Supercontinuum generation in the vacuum ultraviolet through dispersive-wave and soliton-plasma interaction in a noble-gas-filled hollow-core photonic crystal fiber. *Phys. Rev. A* **2015**, *92*, 033821. [CrossRef]
14. Novoa, D.; Cassataro, M.; Travers, J.C.; Russell, P.S.J. Photoionization-Induced Emission of Tunable Few-Cycle Midinfrared Dispersive Waves in Gas-Filled Hollow-Core Photonic Crystal Fibers. *Phys. Rev. Lett.* **2015**, *115*, 033901. [CrossRef] [PubMed]
15. Sollapur, R.; Kartashov, D.; Zürch, M.; Hoffmann, A.; Grigorova, T.; Sauer, G.; Hartung, A.; Schwuchow, A.; Bierlich, J.; Kobelke, J.; et al. Resonance-enhanced multi-octave supercontinuum generation in hollow-core fibers. *Light Sci. Appl.* **2017**, *6*, e17124. [CrossRef]
16. Tani, F.; Köttig, F.; Novoa, D.; Keding, R.; Russell, P.S.J. Effect of anti-crossings with cladding resonances on ultrafast nonlinear dynamics in gas-filled photonic crystal fibers. *Photonics Res.* **2018**, *6*, 84–88. [CrossRef]
17. Meng, F.C.; Liu, B.W.; Wang, S.J.; Liu, J.K.; Li, Y.F.; Wang, C.Y.; Zheltikov, A.M.; Hu, M.L. Controllable two-color dispersive wave generation in argon-filled hypocycloid-core kagome fiber. *Opt. Express* **2017**, *25*, 32972–32984. [CrossRef]
18. Huang, Z.Y.; Chen, Y.F.; Yu, F.; Wang, D.; Zhao, R.R.; Zhao, Y.; Gao, S.F.; Wang, Y.Y.; Wang, P.; Pang, M.; et al. Continuously wavelength-tunable blueshifting soliton generated in gas-filled photonic crystal fibers. *Opt. Lett.* **2019**, *44*, 1805–1808. [CrossRef]
19. Chen, Y.F.; Huang, Z.Y.; Yu, F.; Wu, D.K.; Fu, J.H.; Wang, D.; Pang, M.; Leng, Y.X.; Xu, Z.Z. Photoionization-assisted, high-efficiency emission of a dispersive wave in gas-filled hollow-core photonic crystal fibers. *Opt. Express* **2020**, *28*, 17076–17085. [CrossRef]
20. Hölzer, P.; Chang, W.; Travers, J.C.; Nazarkin, A.; Nold, J.; Joly, N.Y.; Saleh, M.F.; Biancalana, F.; Russell, P.S.J. Femtosecond Nonlinear Fiber Optics in the Ionization Regime. *Phys. Rev. Lett.* **2011**, *107*, 203901. [CrossRef]
21. Huang, Z.Y.; Chen, Y.F.; Yu, F.; Wu, D.K.; Zhao, Y.; Wang, D.; Leng, Y.X. Ionization-induced adiabatic soliton compression in gas-filled hollow-core photonic crystal fibers. *Opt. Lett.* **2019**, *44*, 5562–5565. [CrossRef] [PubMed]
22. Bache, M.; Habib, M.S.; Markos, C.; Lægsgaard, J. Poor-man's model of hollow-core anti-resonant fibers. *J. Opt. Soc. Am. B* **2019**, *36*, 69–80. [CrossRef]
23. Zeisberger, M.; Schmidt, M. Analytic model for the complex effective index of the leaky modes of tube-type anti-resonant hollow core fibers. *Sci. Rep.* **2017**, *7*, 11761. [CrossRef] [PubMed]
24. Chang, W.; Nazarkin, A.; Travers, J.C.; Nold, J.; Hölzer, P.; Joly, N.Y.; Russell, P.S.J. Influence of ionization on ultrafast gas-based nonlinear fiber optics. *Opt. Express* **2011**, *19*, 21018–21027. [CrossRef] [PubMed]
25. Huang, Z.Y.; Wang, D.; Chen, Y.F.; Zhao, R.R.; Zhao, Y.; Nam, S.; Lim, C.; Peng, Y.J.; Du, J.; Leng, Y.X. Wavelength-tunable few-cycle pulses in visible region generated through soliton-plasma interactions. *Opt. Express* **2018**, *26*, 34977–34993. [CrossRef] [PubMed]

26. Huang, Z.Y.; Chen, Y.F.; Yu, F.; Wu, D.K.; Wang, D.; Zhao, R.R.; Zhao, Y.; Gao, S.F.; Wang, Y.Y.; Wang, P.; et al. Highly-tunable, visible ultrashort pulses generation by soliton-plasma interactions in gas-filled single-ring photonic crystal fibers. *Opt. Express* **2019**, *27*, 30798–30809. [CrossRef] [PubMed]
27. Perelomov, A.M.; Popov, V.S.; Terent'ev, M.V. Ionization of atoms in an alternating electric field. *Sov. Phys. JETP* **1966**, *23*, 924–934.
28. Skryabin, D.V.; Luan, F.; Knight, J.C.; Russell, P.S.J. Soliton Self-Frequency Shift Cancellation in Photonic Crystal Fibers. *Science* **2003**, *301*, 1705–1708. [CrossRef]

Article

Understanding Nonlinear Pulse Propagation in Liquid Strand-Based Photonic Bandgap Fibers

Xue Qi [1], Kay Schaarschmidt [1,2], Guangrui Li [1], Saher Junaid [1], Ramona Scheibinger [1,2], Tilman Lühder [1] and Markus A. Schmidt [1,2,3,*]

1 Leibniz Institute of Photonic Technology, Albert-Einstein-Str. 9, 07745 Jena, Germany; xue.qi@leibniz-ipht.de (X.Q.); kay.schaarschmidt@leibniz-ipht.de (K.S.); guangrui.li@uni-jena.de (G.L.); saher.junaid@leibniz-ipht.de (S.J.); ramona.scheibinger@leibniz-ipht.de (R.S.); tilman.luehder@leibniz-ipht.de (T.L.)

2 Abbe Center of Photonics, Faculty of Physics, Friedrich-Schiller-University Jena, Max-Wien-Platz 1, 07743 Jena, Germany

3 Otto Schott Institute of Materials Research (OSIM), Friedrich Schiller University Jena, Fraunhoferstr. 6, 07743 Jena, Germany

* Correspondence: markus-alexander.schmidt@uni-jena.de

Abstract: Ultrafast supercontinuum generation crucially depends on the dispersive properties of the underlying waveguide. This strong dependency allows for tailoring nonlinear frequency conversion and is particularly relevant in the context of waveguides that include geometry-induced resonances. Here, we experimentally uncovered the impact of the relative spectral distance between the pump and the bandgap edge on the supercontinuum generation and in particular on the dispersive wave formation on the example of a liquid strand-based photonic bandgap fiber. In contrast to its air-hole-based counterpart, a bandgap fiber shows a dispersion landscape that varies greatly with wavelength. Particularly due to the strong dispersion variation close to the bandgap edges, nanometer adjustments of the pump wavelength result in a dramatic change of the dispersive wave generation (wavelength and threshold). Phase-matching considerations confirm these observations, additionally revealing the relevance of third order dispersion for interband energy transfer. The present study provides additional insights into the nonlinear frequency conversion of resonance-enhanced waveguide systems which will be relevant for both understanding nonlinear processes as well as for tailoring the spectral output of nonlinear fiber sources.

Keywords: photonic bandgap fiber; dispersive wave; resonance; dispersion management; supercontinuum generation

Citation: Qi, X.; Schaarschmidt, K.; Li, G.; Junaid, S.; Scheibinger, R.; Lühder, T.; Schmidt, M.A. Understanding Nonlinear Pulse Propagation in Liquid Strand-Based Photonic Bandgap Fibers. *Crystals* **2021**, *11*, 305. https://doi.org/10.3390/cryst11030305

Academic Editor: David Novoa

Received: 28 February 2021
Accepted: 17 March 2021
Published: 19 March 2021

1. Introduction

Photonic crystal fibers (PCFs) [1,2] represent a sophisticated type of silica microstructured optical fibers that consist of air holes running along the fiber axis and are widely used for supercontinuum generation (SCG) [3–5]. In case the cylindrical inclusions forming the periodic lattice in the cladding include a material with a refractive index higher than that of glass (established by fluids [6,7], or solids [8,9]) a photonic bandgap (PBG) structure can form, which allows for light guidance in the low index core in selected spectral domains, which are subsequently referred to as transmission bands. The wavelengths corresponding to minimal core mode transmission (i.e., high loss peaks), named here as resonance wavelength, can be estimated by the condition [10]:

$$\lambda_m = 2 \times D_{strand} / (m + 0.5) \times (n_{liquid}^2 - n_{core}^2)^{0.5} \ (m = 1,2,3, \dots) \quad (1)$$

Here, D_{strand} is the diameter of the cylindrical strands. Due to the strong wavelength dependence of this interference effect, PBG fibers (PBGFs) are highly suitable for the spectral filtering of multiple selected wavelengths, with extinction ratios as high as 60 dB/cm [9], or for ultrahigh sensitivity sensors to measure temperature [11] or strain [12]. In addition, the inclusion of resonances into the waveguiding system imposes a unique dispersion landscape [13] that strongly differs from its non-resonant counterpart [14,15] and that for instance leads to a massive increase in the group velocity dispersion (GVD) in close proximity to the edge of the transmission bands. This allowed for the observation of sophisticated nonlinear optical effects in PBGFs, examples of which include the extreme deceleration of the soliton self-frequency shift (SSFS) or the cancellation of the SSFS at the long-wavelength edge of the bandgap regions [16,17]. Within this context another interesting phenomenon is the phase-matched dispersive wave (DW) generation across high attenuation regions between adjacent transmission bands, which for instance has been recently observed in gas-filled anti-resonant fibers [18]. In PBGFs established by high-index fluids [19,20] or high-index germanium-doped inclusions [21], it has been experimentally shown that it is not possible to extend the supercontinuum (SC) bandwidth beyond the bandgap, irrespective of the pump wavelength or power. In a hybrid PCF [22,23] combining the total internal reflection and the bandgap effect, however, DWs can be generated across the resonance in a different transmission band, which was numerically demonstrated [24]. Despite the great success of the reported results, there has been no detailed study on how the wavelength difference between the pump and resonance changes the SCG in PBGFs.

In this work, the impact of the relative spectral distance between the pump and resonance on nonlinear pulse propagation was experimentally demonstrated by using a carbon disulfide (CS_2)-based PBGF. Moreover, the influence of third order dispersion on energy transfer between different bandgaps is discussed. This experimental result and discussion could be used for PBGF design to develop characteristic wavelength converters and enhance the flexibility of SCG in the future.

2. Optical Properties of Liquid Strand Bandgap Fiber

The PBGF used in this work was established by a regular air-hole-based endlessly single mode PCF that has been filled by CS_2, which has a refractive index substantially higher than that of silica. The used PCF (LMA-5, NKT photonics A/S, Blokken, Birkerød, Denmark, Figure 1a) has a pure silica core (core diameter approx. 5.4 µm, refractive index 1.46 @ 0.7 µm) surrounded by six rings of CS_2 strands (diameter D_{strand} = 1.44 µm, pitch Λ = 3.6 µm, refractive index 1.61 @ 0.7 µm) that are arranged in a hexagonal lattice with the central strand omitted, forming the defect core region. Since the CS_2-filled strands in the cladding have a higher refractive index than the silica core, the fiber shows the PBG effect, i.e., yields spectral regions of high and low transmission. Three PBGs can be identified in the range of 500 nm < λ < 1200 nm from the measured loss spectrum of the fundamental core mode (Figure 1b, filled area with green, cyan and yellow colors). The modal loss of the core mode is determined through spectrally monitoring the filling process, i.e., by selectively taking output spectra during filling process, defining a cutback-like method. Specifically, white light from a broadband source (450–2400 nm, SuperK COMPACT, NKT photonics, Blokken, Birkerød, Denmark) is launched into the empty fiber (length 12 cm) mounted onto optofluidic mounts on both sides, and the transmitted signal is fed to an optical spectrum analyzer (OSA). After coupling and optimizing the transmission through the empty fiber, the optofluidic mount on the output side is filled with CS_2 leading to dynamically extending PBG-section starting at the output side with the bandgap loss becoming more and more significant. During the filling process, selected transmitted spectra are acquired via a LabVIEW (National Instruments, Austin, TX, USA) routine. Taking into account the well-known viscosity of CS_2 [25] and the strand diameter (D_{strand} = 1.44 µm), Washburn's equation [26] allows determining the filling length from the file's timestamp. As a consequence, the acquired spectra can be correlated to a particular length of the PBG section and the modal loss can be retrieved via a linear fitting of the transmission (in log scale)/filled length dependency for each wavelength. Here, the data (Figure 1b) are smoothed by a 10 nm moving average for better visualization. The propagation loss of the PCF is neglected here and the scanning speed of the grating of the OSA is not accounted for.

In our nonlinear experiment, the pump laser wavelength (695–710 nm) located in the second order PBG and its long-wavelength edge (−5 dB) at 715 nm is named λ_e in the following. As mentioned in the introduction, the resonant nature of the waveguiding mechanism in PBGFs gives rise to very sophisticated dispersion properties. The dispersion of the fundamental core mode (examples of mode patterns within and outside the transmission bands are given in Figure 1c,d) has been simulated using finite-element modeling (COMSOL Multiphysics, COMSOL Inc., Burlington, MA, USA) including material dispersion but neglecting material loss for both silica and CS_2 (the material dispersion of CS_2 has been taken from Reference [27]). Figure 1b shows the spectral distribution of the GVD parameter $\beta_2 = \partial^2\beta/\partial\omega^2$ (with the propagation constant $\beta = n_{eff}(\omega) \times \omega/c$, effective refractive index n_{eff}, angular frequency ω, and speed of light in vacuum c) in three transmission bands, which is essential for the SCG process and shows a characteristic evolution within each band with a different sign of β_2. This type of behavior is in qualitative accordance with the dispersion of anti-resonance hollow-core fibers [15,28] suggesting a conceptual similarity that results from the general concept of including resonances into the waveguide system for dispersion manipulation, which was also recently demonstrated on the example of dielectric nano-films located on the core of an exposed core fiber [13]. In the present case, the zero dispersion wavelengths (ZDWs) are 553, 689 and 962 nm within these three transmission bands.

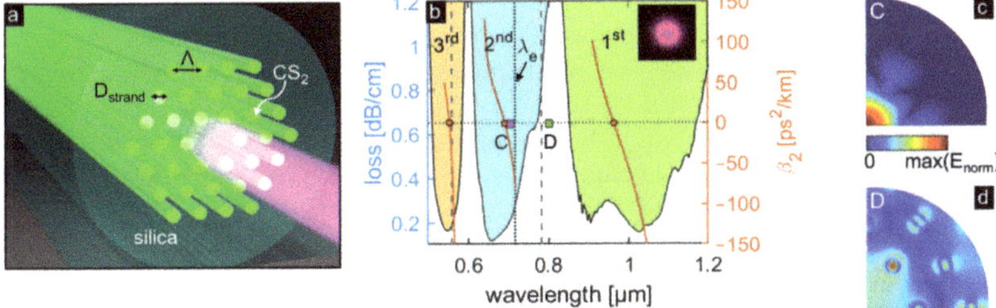

Figure 1. (**a**) Sketch of the CS_2- photonic bandgap fiber (PBGF) (green: hexagonal array of high-index CS_2 strands with diameter D_{strand} = 1.44 μm and pitch Λ = 3.6 μm; semitransparent dark green: silica; pink: light emitted from the silica core); (**b**) measured loss spectrum of the fundamental mode (filled areas with green, cyan and yellow colors, left axis) and the corresponding simulated group velocity dispersion (GVD) parameter β_2 (orange, right axis). The orange dots indicate three zero dispersion wavelengths (ZDWs). The two grey dashed vertical lines indicate the calculated resonance wavelengths from Equation (1) with m = 2 and 3. The black dotted line indicates the bandgap edge (−5 dB) at 715 nm of the second order PBG. The inset in (**b**) is the output beam profile captured by the CCD camera (image extension 265 μm: 50 pixels with each pixel size 5.3 μm). The mode field patterns shown in (**c**,**d**) refer to the simulated mode profiles inside a selected transmission band ((**c**) λ = 700 nm, purple square in (**b**) labeled C) and inside a resonance region ((**d**) λ = 800 nm, green square in (**b**) labeled D).

3. Nonlinear Experiments—DW Generation

The experimental setup used for the nonlinear experiments is shown in Figure 2. Here, a wavelength-tunable Ti:Sapphire laser system (Chameleon-VIS laser, Coherent, Santa Clara, CA, USA, repetition rate 80 MHz, pulse duration 100 fs) acted as an excitation laser allowing for adjusting the central pump wavelength between 695 nm $< \lambda_P <$ 710 nm and thus enabling the investigation of the SCG process as a function of spectral distance between the pump and bandgap edge of PBG $\Delta\lambda = \lambda_e - \lambda_P$. Note that even though the tuning range of the laser is starting at 690 nm, the laser does not stably operates at this wavelength, thus the investigation starts at 695 nm. Via a combination of a thin-film polarizer and a half-wave plate inserted after the polarized pump laser, the input power launched into the fiber was controlled. The beam was coupled into the core area of the CS_2-PBGF (length 0.5 m) using an anti-reflection coated lens (A375-B, Thorlabs, Newton, NJ, USA, focal length 7.5 cm, NA 0.3). A transmission efficiency of 40% (ratio of average output power from the objective and input power before the lens) was achieved. This value was kept constant for different pump wavelengths by slightly adjusting the position of the input lens. The output beam from the fiber was collimated by a 10× objective and guided by a multimode fiber to the spectrometer (AQ-6315A, Ando Electric, Kanagawa, Japan). To monitor the pattern of the mode, a small fraction of the beam was reflected by a wedge and monitored by a CCD camera (Thorlabs, Newton, NJ, USA).

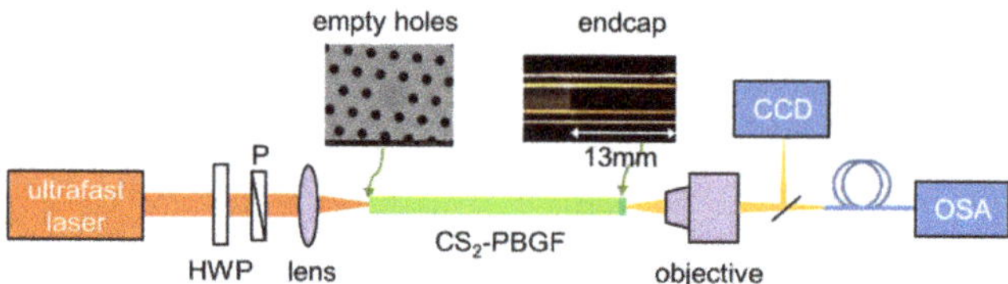

Figure 2. Sketch of the ultrafast experimental setup to characterize the supercontinuum generation process in the CS_2-PBGF (HWP: half-wave plate; P: polarizer; OSA: optical spectrum analyzer). The left top image is the SEM picture of the used PCF without fluid inside. The right top image shows a microscopic image of the spliced endcap (the green part at the end of the CS_2-PBGF, details in the main text).

Note that the input side of the CS_2-PBGF remains open to the environment, while the output side is blocked by a home-made endcap. The endcap is a short piece of single mode fiber (780-HP, Nufern, CT, USA, length 13 mm), spliced to the output side of the CS_2-PBGF to prevent the liquid from evaporating. The reason why there is no endcap at the input side is that the damage of the fiber was observed at low input power in the experiment when the CS_2-PBGF was blocked by endcaps at both sides. We also want to stress that mounting the fiber into optofluidic mounts on both sides [29] was not appropriate for this case. After exposure to short pulses, some debris was observed on the surface of the solid core, which caused the coupling efficiency to dramatically drop when the input power was increased. We attribute this damage to the liquid–solid interface within the focal spot: the liquid burns more easily on this interface compared to liquid–core fibers which show only a transition from bulk liquid to liquid core. Therefore, from the practical perspective, we leave the input side of the fiber open to the environment to prevent a liquid–solid interface. This increases the damage threshold of the CS_2-PBGF at the expense of fluid evaporation. The evaporating rate of CS_2 is measured and the effective sample length is calibrated accordingly (see Appendix A Figure A1 for details). Although the damage threshold can be increased effectively using the present method (the input side left free and the output side blocked by an endcap), fiber damage is still observed at peak powers around 2.5 kW (pulse energy 0.25 nJ).

In our previous numerical simulation work [30], we revealed that the spectral distance between our pump and resonance plays a crucial role in the SCG process of PBGF. In the present work, this impact is studied from the experimental perspective. Figure 3a–e show the nonlinear dynamics of SCG process with an increasing pulse energy for different spectral distances $\Delta\lambda = \lambda_e - \lambda_p$ between the pump λ_p and the bandgap edge λ_e of second order PBG. The spectral distance $\Delta\lambda$ has important influence on the formation of a DW, which is emitted in the normal dispersion (ND) regime by a soliton that is formed from the input pulse after initial self-phase modulation (SPM). When $\Delta\lambda$ is decreased from 20 to 10 nm (Figure 3a–d), the generated DW blue shifts (spectral distance between the pump and DW ($\Delta\lambda_{DW} = \lambda_p - \lambda_{DW}$) increases). This is accompanied by a higher threshold for DW generation (Figure 3f), which is associated with the increasing spectral distance between soliton and DW [31]. For the case $\Delta\lambda$ = 5 nm, the pump energy is not high enough to generate DW radiation. These experimental results can be supported by simulations of nonlinear pulse propagation based on the generalized nonlinear Schrödinger equation (GNLSE) [32] (see Appendix A Figure A2 for details).

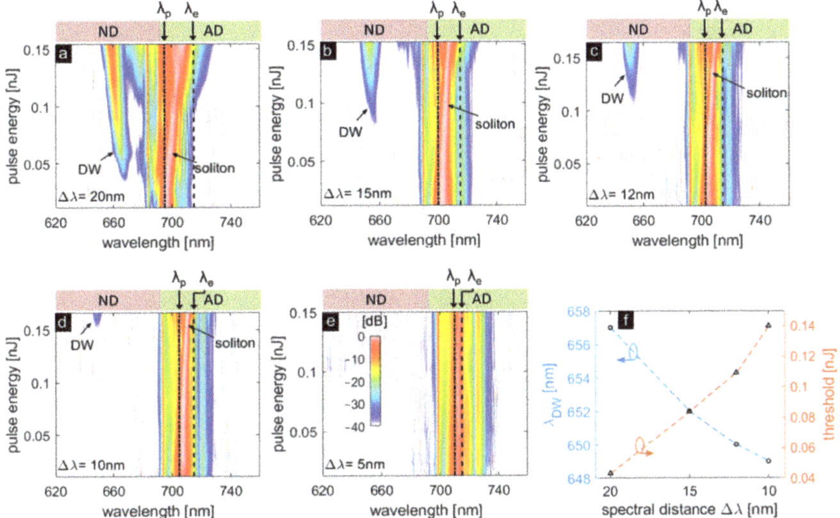

Figure 3. (**a–e**) Measured energy/spectral evolutions of the supercontinuum generation (SCG) process in the CS_2-PBGF for various spectral distances $\Delta\lambda = \lambda_e - \lambda_p$ between the pump λ_p and bandgap edge λ_e of second order PBG ((**a–e**) $\Delta\lambda$ = 20, 15, 12, 10 and 5 nm). The black dashed line indicates λ_e and the black dash-dotted line indicates λ_p; (**f**) dispersive wave (DW) wavelength λ_{DW} and threshold of the generated DW both as functions of $\Delta\lambda$.

The significantly different nonlinear pulse evolution for various values of $\Delta\lambda$ is caused by the strongly changed dispersion landscape in the vicinity of the bandgap edge. The behavior of DW can be explained particularly by the phase-matching (PM) condition of soliton and DW as outlined in the following. DW generation results from the fission of a higher order soliton into its fundamental counterparts, which release the excess energy to DW. DW generation is most effective at the wavelength that fulfills the PM condition defined by the following equation [4,31]:

$$\Delta\beta_{DW}(\omega) = \beta(\omega) - [\beta(\omega_{sol}) + (\omega - \omega_{sol})\beta_{1,sol} + \frac{1}{2} \times \gamma \times P_{sol}] \quad (2)$$

Here, ω and ω_{sol} are the frequencies of DW and soliton, respectively, $\beta_{1,sol} = \partial\beta/\partial\omega\,|\,\omega_{sol}$ is the inverse group velocity at ω_{sol}, P_{sol} is the peak power of the assumed first soliton (with input peak power P_0 and soliton number N, $P_{sol} = (2 - 1/N)^2 \times P_0$) [31], and $\gamma = \omega_{sol} \times n_2/c/A_{eff}$ is the nonlinear coefficient (A_{eff} is the effective mode area and n_2 is the nonlinear refractive index). The condition $\Delta\beta_{DW} = 0$ determines the phase-matched spectral location of DW (λ_{DW}) in the case the soliton wavelength (λ_{sol}) is given or vice versa. Note that in the experiments the observed λ_{sol} is located in close proximity to or inside the resonant (high-loss) region and thus this wavelength cannot be unambiguously determined, whilst λ_{DW} is clearly visible in the energy/spectral evolutions (Figure 3a–d). Therefore, the λ_{sol} is calculated from the measured spectral position and the threshold of DW (taken from its creation position, 667 nm at 0.04 nJ, 657 nm at 0.08 nJ, 654 nm at 0.11 nJ, and 649 nm at 0.15 nJ corresponding to Figure 3a–d, respectively) via Equation (2) and is shown in Figure 4. It can be seen that the solitons are located between 704 nm < λ < 713 nm (marked as gray bar in Figure 4) for the cases of $\Delta\lambda$ = 20, 15, 12 and 10 nm, which agrees well with the experimental results in Figure 3a–e. It should be noted that in the dispersion simulation, a small deviation in the hole diameter (2%) will cause a 10 nm deviation of ZDW and thus induce a significant discrepancy from the experimental results (see Appendix A Figure A3 and Table A1 for details). While such structural asymmetries may only have a minor effect on the transmission spectra of anti-resonant reflecting optical waveguide (ARROW) PCFs, it is quite likely that they greatly affect the higher-order dispersion terms that are important for ultrafast nonlinear frequency conversion.

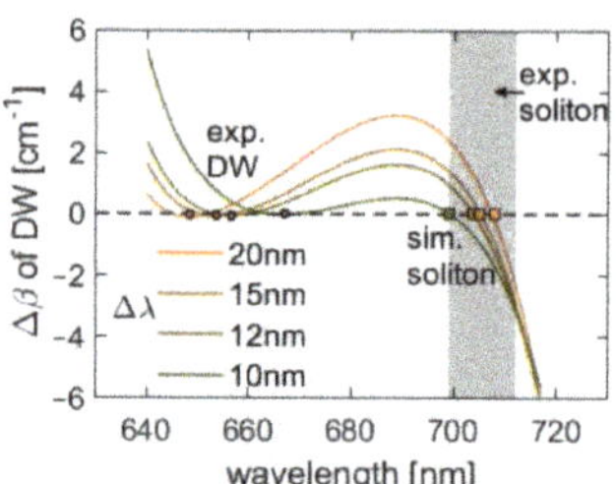

Figure 4. The calculated phase-mismatching rate of DW generation $\Delta\beta_{DW}$ in the same second order PBG as the pump for different spectral distances $\Delta\lambda$ (γ = 6.5 W^{-1}/km, pulse: 100 fs, 80 MHz) using Equation (2). The wavelengths of DWs are taken from the experiments and the P_{sol} is calculated from the DW generation threshold. The corresponding soliton wavelengths are obtained by the crossing of the individual curves with the zero line ($\Delta\beta$ = 0), corresponding to DW-PM. The gray area means the soliton wavelengths taken from experiment.

To clearly demonstrate that filling the PCF with CS_2 leads to a massive change of waveguide dispersion and thus to the nonlinear frequency conversion, Figure 5 shows the measured output spectra of the CS_2-PBGF (blue) in comparison to the same PCF with air holes (orange) both at the same pump peak power (1.5 kW) and pump wavelength (λ_p = 700 nm). For the empty PCF, only SPM in close proximity to λ_p can be observed, whilst efficient DW generation is visible for the CS_2-PBGF due to the entirely different dispersion landscape. Note that the ZDW of the air-hole based PCF is ~1400 nm, which is an unsuitable scenario for DW generation pumped at around 700 nm. The CS_2-PBGF, however, shows a ZDW at 689 nm which is induced by the inclusion of the resonance and enables DW formation. The sharp spectral peak at around 690 nm which can be seen in Figures 3a–e and 5 is a feature of the pump laser.

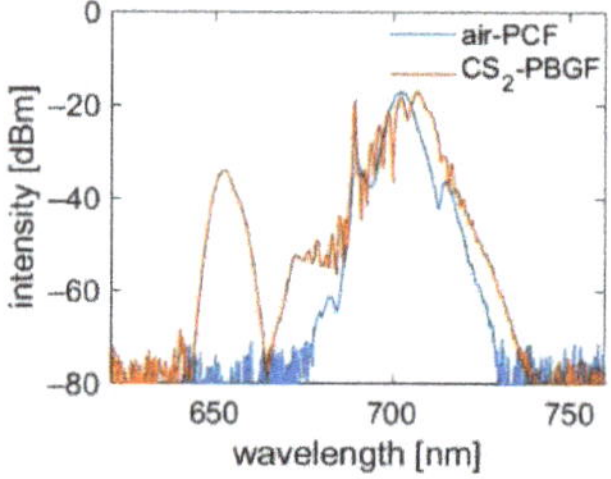

Figure 5. Comparison of the measured nonlinear frequency conversion spectra in the CS_2-PBGF (blue) and air-hole based photonic crystal fibers (PCF) (orange) both for the same pump peak power (1.5 kW) and pump wavelength (700 nm).

4. Discussion

To investigate the critical parameter that determines whether DWs can be generated across the resonances (i.e., high attenuation regions) between different PBGs, the propagation constant $\beta(\omega)$ is expanded into a Taylor series around the soliton frequency ω_{sol} and is inserted into the DW-PM condition (Equation (2)). We neglect the nonlinear phase which only plays a minor role here because of the low pulse power (confirmed in Figure 6), leading to [23]:

$$\Delta\beta_{DW}(\lambda) = 2 \times (\pi \times c/\lambda_{sol})^2 \times (\lambda_{sol}/\lambda - 1)^2 \times \beta_2 + 4/3 \times (\pi \times c/\lambda_{sol})^3 \times (\lambda_{sol}/\lambda - 1)^3 \times \beta_3 \quad (3)$$

where $\beta_2 = \partial^2\beta/\partial\omega^2 \mid_{\omega sol}$ and $\beta_3 = \partial^3\beta/\partial\omega^3 \mid_{\omega sol}$ are the group-velocity dispersion and third order dispersion at ω_{sol}. Therefore, the estimated dispersive wave λ_{DW}, defined by the condition $\Delta\beta_{DW}(\lambda_{DW}) = 0$, can be written as [23]

$$\lambda_{DW} = \lambda_{sol}/[1 - \beta_2 \times \lambda_{sol}/(2 \times \pi \times c \times \beta_3)] \quad (4)$$

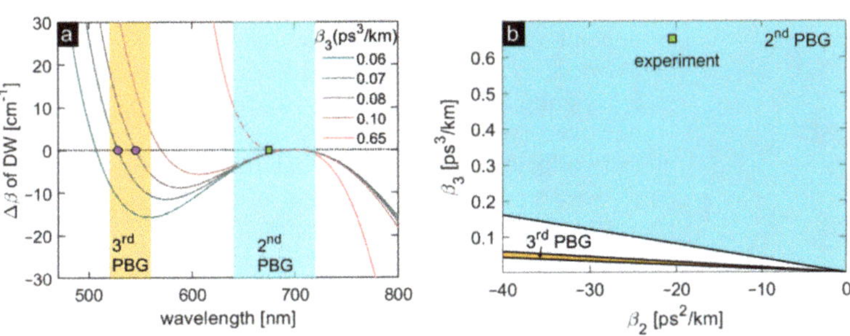

Figure 6. (**a**) Analytical description of the phase-mismatching rate $\Delta\beta_{DW}$ as a function of β_3 in the second order PBG with $\beta_2 = -20.4$ ps^2/km corresponding to $\Delta\lambda = 20$ nm case. The solid lines indicate $\Delta\beta_{DW}$ neglecting the nonlinear phase (Equation (3)) while the dashed lines indicate $\Delta\beta_{DW}$ with the nonlinear phase (Equation (2), $\gamma = 6.5$ W^{-1}/km, N = 1.6, $P_{sol} = 1$ kW corresponding to DW generation threshold in Figure 3b). The cyan bar indicates the second order PBG and the yellow bar indicates the third order PBG of the sample. The green square corresponds to the $\Delta\lambda = 20$ nm case in Figure 3b. The purple dots indicate the DW generated across the resonance band; (**b**) the estimated DW wavelength for different combinations of β_2 and β_3 from Equation (4) (soliton wavelength $\lambda_{sol} = 699$ nm corresponding to the $\Delta\lambda = 20$ nm case). The cyan area indicates the second order PBG and the yellow area indicates third order PBG of the sample. The green square indicates the generated DW in the experiment for the $\Delta\lambda = 20$ nm case.

To demonstrate the critical impact of β_3 and its interplay with β_2 on the interband DW-generation, the DW-PM condition (including Equation (2) and the simplified one Equation (3)) and estimated DW wavelength from Equation (4) are shown in Figure 6, taking the parameters of $\Delta\lambda = 20$ nm case (soliton wavelength $\lambda_{sol} = 699$ nm). Figure 6a shows the DW-PM condition for different values of β_3 with fixed $\beta_2 = -20.4$ ps^2/km corresponding to $\Delta\lambda = 20$ nm case. Only when β_3 is within the range of 0.06–0.1 ps^3/km, the PM condition can be satisfied for interband DW-generation. This can explain why no interband DW-generation was detected in the experiment ($\beta_3 = 0.65$ ps^3/km, see Table 1; the DW generated in the experiment is indicated as the green square in Figure 6a). To investigate the impact of β_3 and its interplay with β_2 on the interband DW-generation further and give some insights into the fiber design in the future to realize the interband DW-generation, the estimated DW for the different combination of β_2 and β_3 from Equation (4) is plotted in Figure 6b (cyan area: second order PBG; yellow area: third order PBG). It can be seen that only a small part of combinations of β_2 and β_3 can satisfy the PM condition to realize the interband energy transfer from the second to third PBG. The situation for the experiment conducted in this work (dispersion parameters β_2 and β_3 of the sample) is shown in Table 1 and for the $\Delta\lambda = 20$ nm case, the DW is indicated as a green square in Figure 6b, revealing that the established combinations of β_2 and β_3 prevent the interband DW-generation.

Table 1. Dispersion parameters β_2 and β_3 ($\beta(\omega)$ come from the numerical simulation (COMSOL)) at soliton wavelengths (taken from the experiments, Figure 3a–e) for different spectral distances $\Delta\lambda$.

$\Delta\lambda$	20 nm	15 nm	12 nm	10 nm	5 nm
λ_{sol} (nm)	699	704	708	710	712
β_2 (ps^2/km)	−20.4	−33.6	−45.8	−52.3	−59.9
β_3 (ps^3/km)	0.65	0.76	1.00	1.13	1.15

5. Conclusions

Understanding the impact of dispersion on nonlinear frequency conversion is particularly relevant within the context of ultrafast supercontinuum generation, which crucially depends on the dispersive properties of the underlying waveguide. This impact is particularly strong for optical waveguides that include geometry-induced resonances, leading to a significant modification of higher-order dispersion terms. Here, we experimentally unlocked the impact of the relative spectral distance between the pump and resonance on nonlinear pulse propagation through the example of a liquid-based PBGF. The PBGF shows a sophisticated dispersive landscape leading to dispersive wave generation on the short-wavelength side of the pump. Due to the dramatic change of the dispersion in close proximity to the bandgap edges, an adjustment of the pump wavelength by a few nanometers results in a different nonlinear pulse propagation situation, i.e., in the dispersive wave to be generated at a different wavelength. This behavior is explained by phase-matching considerations between the soliton and DW, additionally revealing the critical influence of third order dispersion for interband energy transfer. The present study gives additional insights into nonlinear frequency conversion of resonance-enhanced waveguide systems which will be relevant for both understanding nonlinear processes as well as for tailoring the spectral output of nonlinear fiber sources.

Author Contributions: Conceptualization, M.A.S.; methodology, X.Q., K.S., G.L. and M.A.S.; software, X.Q. and T.L.; validation, X.Q. and M.A.S.; formal analysis, X.Q. and M.A.S.; investigation, X.Q.; resources, K.S., S.J. and R.S.; data curation, X.Q.; writing—original draft preparation, X.Q.; writing—review and editing, K.S., S.J., R.S., T.L. and M.A.S.; visualization, X.Q., K.S. and T.L.; supervision, M.A.S.; project administration, M.A.S.; funding acquisition, M.A.S. All authors have read and agreed to the published version of the manuscript.

Funding: This research was funded by Deutsche Forschungsgemeinschaft, grant number SCHM2655/11-1, SCHM2655/12-1, SCHM2655/8-1, 259607349/GRK2101; H2020 Marie Skłodowska-Curie Actions, grant number 713694.

Institutional Review Board Statement: Not applicable.

Informed Consent Statement: Not applicable.

Data Availability Statement: The data presented in this study are available from the corresponding author on reasonable request.

Acknowledgments: The authors thank Tomas Cizmar and Sergey Turtaev for the support of Ti:Sapphire laser system. The authors acknowledge support by the German Research Foundation and the Open Access Publication Fund of the Thueringer Universitaets- und Landesbibliothek Jena Projekt-Nr. 433052568.

Conflicts of Interest: The authors declare no conflict of interest.

Appendix A

The evaporation rate of CS_2 in PCF (NKT-5) at room temperature is measured with the help of an optical microscope (the other side of PCF is closed with an endcap, PCF length 40 cm). As shown in Figure A1, the evaporation rate is high at the beginning but drops dramatically after 50 min and stabilizes at 0.8 μm/s in 2 h. At the same time, the measured fiber length, in which CS_2 has been evaporated, increases rapidly at first and then increases linearly at a lower rate. Since the nonlinear experiment in this paper is implemented in 2 h, the effective sample length can be calibrated by the original length minus 1 cm.

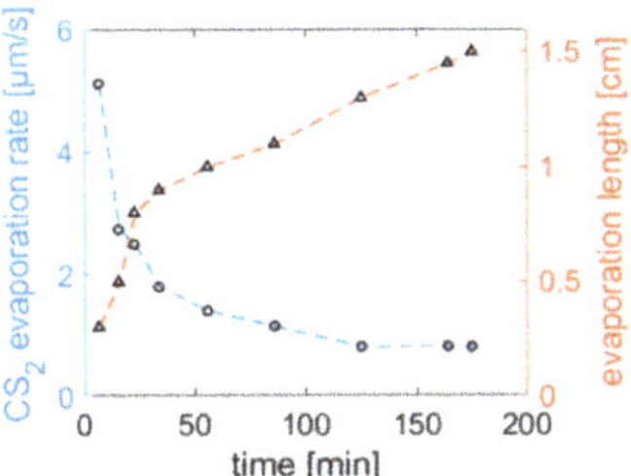

Figure A1. Measured evaporation rate of CS_2 in PCF (blue line) and the length of fiber in which CS_2 has been evaporated (red line).

The simulated energy/spectral evolutions of the SCG process in the CS_2-PBGF with D_{strand} = 1.44 μm for various spectral distances $\Delta\lambda = \lambda_e - \lambda_p$ between the pump λ_p and the bandgap edge λ_e of second order PBG ($\Delta\lambda$ = 20, 15, 12, 10 and 5 nm) are shown in Figure A2a–e. The simulation results show great consistency with experimental ones (Figure 3a–e). These simulations of nonlinear pulse propagation are implemented based on the generalized nonlinear Schrödinger equation (GNLSE) [32] with the parameters: nonlinear coefficient $\gamma = 6.5\ W^{-1}/km$, pulse: 100 fs, 80 MHz, sample length L = 0.5 m. Note that the measured loss (Figure 1b) is included in the simulation.

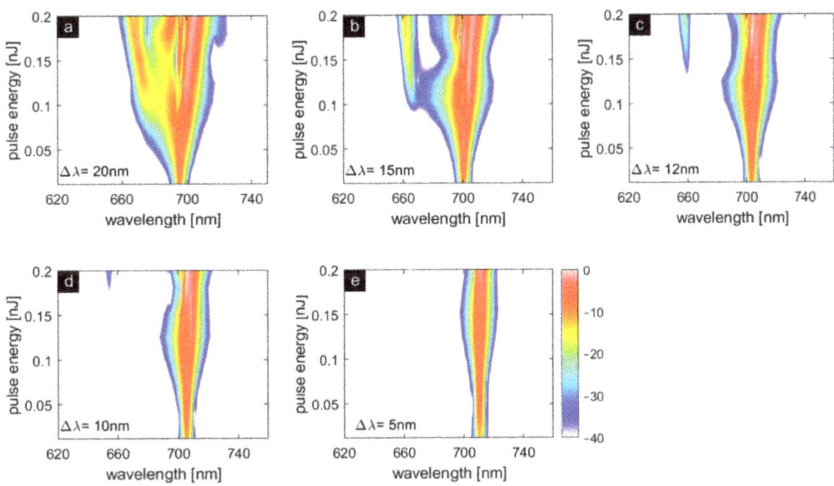

Figure A2. (**a**–**e**) Simulated energy/spectral evolutions of the SCG process in the CS_2-PBGF with D_{strand}= 1.44 μm for various spectral distances $\Delta\lambda = \lambda_e - \lambda_p$ between the pump λ_p and bandgap edge λ_e of second order PBG ($\Delta\lambda$ = 20, 15, 12, 10 and 5 nm). Parameters used: $\gamma = 6.5\ W^{-1}/km$, pulse: 100 fs, 80 MHz, sample length L = 0.5 m.

More details about the measurement error of strand diameters and the corresponding impact on the difference between the simulation and experimental results are discussed in the following. The measured diameter distribution of the air holes in the inner ring of the fiber cladding ranges from 1.462 to 1.497 μm, which we believe results from (i) measurement errors in the SEM imaging (the typical deviation in measuring lengths with our SEM instrument is at least 3%) and (ii) fabrication imperfections. It should be noted that a small deviation in strand diameter (e.g., 2%) will cause a strong change of ZDW and calculated soliton wavelength according to the DW phase-mismatching rate (Equation (2)), which are shown in Figure A3. To determine an average hole diameter required for the numerical calculations, we compared the results of nonlinear pulse propagation simulations for different air hole diameters with the experimental results, revealing that when D_{strand} = 1.44 μm, simulations (Figures 4, A2 and A3b) match well with the experimental results.

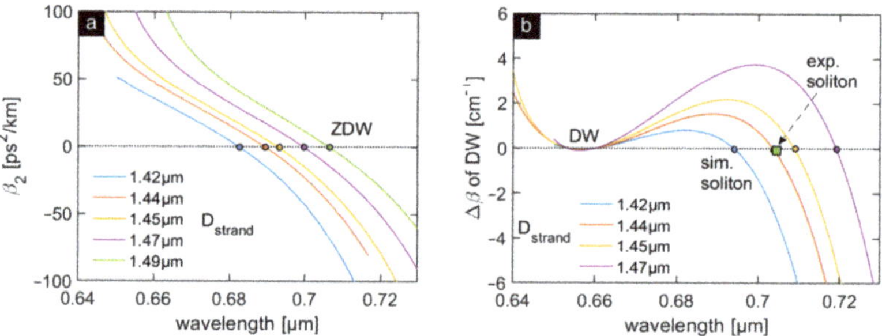

Figure A3. (**a**) Simulated GVD parameter β_2 and (**b**) calculated phase-mismatching rate of DW generation $\Delta\beta_{DW}$ in the second order PBG for different strand diameters (γ = 6.5 W^{-1}/km, pulse: 100 fs, 80 MHz) using Equation (2) of the manuscript. The wavelength of the DW (657 nm) and the P_{sol} calculated from the DW generation threshold (0.08 nJ) are taken from the experiment (Figure 3b). The corresponding soliton wavelengths are obtained by the crossing of the individual curves with the zero line ($\Delta\beta = 0$). Note that the green square in (**b**) indicates the measured soliton wavelength in the experiment.

Table A1. ZDWs and soliton wavelengths calculated by Equation (2) of CS_2-PBGF assuming different strand diameters (the percentage indicates the deviation from D_{strand} = 1.44 μm). The assumed diameters roughly correspond to the intrinsic range of strand diameter deviation of the PCF used that needs to be taken into account. Note that no soliton wavelength can be calculated for D_{strand} = 1.49 μm as the pump pulse (700 nm) locates in the normal dispersion region.

D_{strand} (μm)	1.42	1.44	1.45	1.47	1.49
Deviation from 1.44 μm	−1.4%	0%	+0.7%	+2.1%	+3.5%
ZDW (nm)	683	689	693	700	706
Calculated λ_{sol} with Equation (2) (nm)	694	704	709	719	-

References

1. Russell, P. Photonic Crystal Fibers. *Science* **2003**, *299*, 358–362. [CrossRef]
2. Birks, T.A.; Knight, J.C.; Russell, P.S.J. Endlessly single-mode photonic crystal fiber. *Opt. Lett.* **1997**, *22*, 961–963. [CrossRef]
3. Husakou, A.V.; Herrmann, J. Supercontinuum Generation of Higher-Order Solitons by Fission in Photonic Crystal Fibers. *Phys. Rev. Lett.* **2001**, *87*, 203901. [CrossRef] [PubMed]
4. Dudley, J.M.; Genty, G.; Coen, S. Supercontinuum Generation in Photonic Crystal Fibre. *Rev. Mod. Phys.* **2006**, *78*, 1135–1184. [CrossRef]
5. Qi, X.; Chen, S.; Li, Z.; Liu, T.; Ou, Y.; Wang, N.; Hou, J. High-power visible-enhanced all-fiber supercontinuum generation in a seven-core photonic crystal fiber pumped at 1016 nm. *Opt. Lett.* **2018**, *43*, 1019. [CrossRef] [PubMed]
6. Steinvurzel, P.; Eggleton, B.J.; de Sterke, C.M.; Steel, M.J. Continuously tunable bandpass filtering using high-index inclusion microstructured optical fibre. *Electron. Lett.* **2005**, *41*, 463. [CrossRef]
7. Bise, R.T.; Windeler, R.; Kranz, K.S.; Kerbage, C.; Eggleton, B.J.; Trevor, D.J. Tunable photonic band gap fiber. In Proceedings of the Optical Fiber Communications Conference, Anaheim, CA, USA, 17–22 March 2002. [CrossRef]
8. Schmidt, M.A.; Granzow, N.; Da, N.; Peng, M.; Wondraczek, L.; Russell, P.S.J. All-solid bandgap guiding in tellurite-filled silica photonic crystal fibers. *Opt. Lett.* **2009**, *34*, 1946. [CrossRef]
9. Granzow, N.; Schmidt, M.A.; Tverjanovich, A.S.; Wondraczek, L.; Russell, P.S.J. Band-gap guidance in chalcogenide-silica photonic crystal fibers. *Opt. Lett.* **2011**, *36*, 2432–2434. [CrossRef]

10. Litchinitser, N.M.; Dunn, S.C.; Steinvurzel, P.E.; Eggleton, B.J.; White, T.P.; McPhedran, R.C.; Martijn de Sterke, C. Application of an ARROW model for designing tunable photonic devices. *Opt. Express* **2004**, *12*, 1540. [CrossRef] [PubMed]
11. Cheng, L.; Wu, J.J.; Hu, X.W.; Peng, J.G.; Yang, L.Y.; Dai, N.L.; Li, J.Y. Ultrahigh Temperature Sensitivity Using Photonic Bandgap Effect in Liquid-Filled Photonic Crystal Fibers. *IEEE Photonics J.* **2017**, *9*, 1–7. [CrossRef]
12. Lin, C.; Wang, Y.; Huang, Y.; Liao, C.; Bai, Z.; Hou, M.; Li, Z.; Wang, A.Y. Liquid modified photonic crystal fiber for simultaneous temperature and strain measurement. *Photonics Res.* **2017**, *5*, 129–133. [CrossRef]
13. Lühder, T.A.K.; Schaarschmidt, K.; Goerke, S.; Schartner, E.P.; Ebendorff-Heidepriem, H.; Schmidt, M.A. Resonance-Induced Dispersion Tuning for Tailoring Nonsolitonic Radiation via Nanofilms in Exposed Core Fibers. *Laser Photonics Rev.* **2020**, 1900418. [CrossRef]
14. Zeisberger, M.; Schmidt, M.A. Analytic model for the complex effective index of the leaky modes of tube-type anti-resonant hollow core fibers. *Sci. Rep.* **2017**, *7*, 1–14. [CrossRef] [PubMed]
15. Zeisberger, M.; Hartung, A.; Schmidt, M. Understanding Dispersion of Revolver-Type Anti-Resonant Hollow Core Fibers. *Fibers* **2018**, *6*, 68. [CrossRef]
16. Bétourné, A.; Kudlinski, A.; Bouwmans, G.; Vanvincq, O.; Mussot, A.; Quiquempois, Y. Control of supercontinuum generation and soliton self-frequency shift in solid-core photonic bandgap fibers. *Opt. Lett.* **2009**, *34*, 3083. [CrossRef] [PubMed]
17. Vanvincq, O.; Kudlinski, A.; Bétourné, A.; Mussot, A.; Quiquempois, Y.; Bouwmans, G. Manipulating the Propagation of Solitons with Solid-Core Photonic Bandgap Fibers. *Int. J. Opt.* **2012**, *2012*, 1–12. [CrossRef]
18. Sollapur, R.; Kartashov, D.; Zürch, M.; Hoffmann, A.; Grigorova, T.; Sauer, G.; Hartung, A.; Schwuchow, A.; Bierlich, J.; Kobelke, J.; et al. Resonance-enhanced multi-octave supercontinuum generation in antiresonant hollow-core fibers. *Light Sci. Appl.* **2017**, *6*, e17124. [CrossRef]
19. Fuerbach, A.; Steinvurzel, P.; Bolger, J.A.; Nulsen, A.; Eggleton, B.J. Nonlinear propagation effects in antiresonant high-index inclusion photonic crystal fibers. *Opt. Lett.* **2005**, *30*, 830. [CrossRef]
20. Fuerbach, A.; Steinvurzel, P.; Bolger, J.A.; Eggleton, B.J. Nonlinear pulse propagation at zero dispersion wavelength in anti-resonant photonic crystal fibers. *Opt. Express* **2005**, *13*, 2977–2987. [CrossRef]
21. Kibler, B.; Martynkien, T.; Szpulak, M.; Finot, C.; Fatome, J.; Wojcik, J.; Urbanczyk, W.; Wabnitz, S. Nonlinear femtosecond pulse propagation in an all-solid photonic bandgap fiber. *Opt. Express* **2009**, *17*, 10393. [CrossRef]
22. Arismar Cerqueira, S., Jr.; Cordeiro, C.M.B.; Biancalana, F.; Roberts, P.J.; Hernandez-Figueroa, H.E.; Cruz, C.H.B. Nonlinear interaction between two different photonic bandgaps of a hybrid photonic crystal fiber. *Opt. Lett.* **2008**, *33*, 2080. [CrossRef]
23. Arismar Cerqueira, S.; do Nascimento, A.R.; Bonomini, I.A.M.; Franco, M.A.R.; Serrão, V.A.; Cordeiro, C.M.B. Strong power transfer between photonic bandgaps of hybrid photonic crystal fibers. *Opt. Fiber Technol.* **2015**, *22*, 36–41. [CrossRef]
24. Pureur, V.; Dudley, J.M. Nonlinear spectral broadening of femtosecond pulses in solid-core photonic bandgap fibers. *Opt. Lett.* **2010**, *35*, 2813. [CrossRef]
25. Laesecke, A.; Muzny, C.D. Reference Correlation for the Viscosity of Carbon Dioxide. *J. Phys. Chem. Ref. Data* **2017**, *46*, 013107. [CrossRef] [PubMed]
26. Washburn, E.W. The Dynamics of Capillary Flow. *Phys. Rev.* **1921**, *17*, 273–283. [CrossRef]
27. Chemnitz, M.; Scheibinger, R.; Gaida, C.; Gebhardt, M.; Stutzki, F.; Pumpe, S.; Kobelke, J.; Tünnermann, A.; Limpert, J.; Schmidt, M.A. Thermodynamic control of soliton dynamics in liquid-core fibers. *Optica* **2018**, *5*, 695. [CrossRef]
28. Grigorova, T.; Sollapur, R.; Hoffmann, A.; Hartung, A.; Schwuchow, A.; Bierlich, J.; Kobelke, J.; Schmidt, M.A.; Spielmann, C. Measurement of the Dispersion of an Antiresonant Hollow Core Fiber. *IEEE Photonics J.* **2018**, *10*. [CrossRef]
29. Chemnitz, M.; Gaida, C.; Gebhardt, M.; Stutzki, F.; Kobelke, J.; Tünnermann, A.; Limpert, J.; Schmidt, M.A. Carbon chloride-core fibers for soliton mediated supercontinuum generation. *Opt. Express* **2018**, *26*, 3221. [CrossRef] [PubMed]
30. Qi, X.; Schaarschmidt, K.; Chemnitz, M.; Schmidt, M.A. Essentials of resonance-enhanced soliton-based supercontinuum generation. *Opt. Express* **2020**, *28*, 2557. [CrossRef]
31. Agrawal, G.P. *Nonlinear Fiber Optics*, 5th ed.; Elsevier: Burlington, MA, USA, 2007; ISBN 9780123970237.
32. Dudley, J.M.; Taylor, R. *Supercontinuum Generation in Optical Fibers*; Cambridge University Press: Cambridge, UK, 2010; ISBN 9780521514804.

Article

Polarization Modulation Instability in Dispersion-Engineered Photonic Crystal Fibers

Abraham Loredo-Trejo [1,2], Antonio Díez [1,2,*], Enrique Silvestre [1,3] and Miguel V. Andrés [1,2]

1 Laboratory of Fiber Optics, ICMUV, Universidad de Valencia, 46100 Valencia, Spain; abraham.loredo@uv.es (A.L.-T.); enrique.silvestre@uv.es (E.S.); miguel.andres@uv.es (M.V.A.)
2 Departamento de Física Aplicada y Electromagnetismo, Universidad de Valencia, 46100 Valencia, Spain
3 Departamento de Óptica, Universidad de Valencia, 46100 Valencia, Spain
* Correspondence: antonio.diez@uv.es

Abstract: Generation of widely spaced polarization modulation instability (PMI) sidebands in a wide collection of photonic crystal fibers (PCF), including liquid-filled PCFs, is reported. The contribution of chromatic dispersion and birefringence to the net linear phase mismatch of PMI is investigated in all-normal dispersion PCFs and in PCFs with one (or two) zero dispersion wavelengths. Large frequency shift sidebands are demonstrated experimentally. Suitable fabrication parameters for air-filled and liquid-filled PCFs are proposed as guidelines for the development of dual-wavelength light sources based on PMI.

Keywords: photonic crystal fiber; polarization modulation instability; ANDi fiber; liquid-filled PCF

Citation: Loredo-Trejo, A.; Díez, A.; Silvestre, E.; Andrés, M.V. Polarization Modulation Instability in Dispersion-Engineered Photonic Crystal Fibers. *Crystals* **2021**, *11*, 365. https://doi.org/10.3390/cryst11040365

Academic Editors: David Novoa and Nicolas Y. Joly

Received: 26 February 2021
Accepted: 29 March 2021
Published: 30 March 2021

1. Introduction

In the last decades, great effort has been applied to take advantage of nonlinearity in optical fibers for the development of new light sources as supercontinuum sources [1,2] or four-wave mixing (FWM)-based light sources. Particularly, solid-core photonic crystal fibers (PCF) [3] are an excellent platform for the generation of nonlinear processes, owing to the large flexibility for fiber chromatic dispersion design, the strong optical density, and the large interaction lengths available. FWM has attractive applications in different areas of interest: parametric amplifiers [4], frequency combs based on cascade-FWM [5], dual-wavelength light sources for special microscopy as coherent anti-Stokes Raman spectroscopy [6,7], and novel nonlinear quantum applications as photon-pair sources [8]. FWM is a flexible, nonlinear parametric process in which two pump photons of different (or identical) frequency interact with a nonlinear material to produce two photons frequency-shifted from the pump, one with higher energy than the pump, called anti-Stokes photon, and another with lower energy, called Stokes photon. The frequency shift obeys the energy conservation relation and phase matching condition [9].

In the most common cases, FWM applications rely on the scalar FWM process, in which the state of polarization of the involved photons is identical. Light sources with specific characteristics of polarization can be of great interest in the area of optics. In this sense, vector–FWM (V–FWM) can be exploited for the development of such light sources. V–FWM is ruled by the interplay of the state of polarization of the involved photons with the nonlinear medium. We find two different types of V–FWM processes, named polarization modulation instability (PMI) and cross-phase modulation instability (XPMI) [10]. XPMI can be generated in fibers with moderated birefringence. Linearly polarized light oriented at 45°, with respect the principal fiber axes, produces two sidebands with orthogonal polarization. PMI is mostly generated in very weakly birefringent fibers, in which a linearly polarized optical pump signal with polarization oriented along one fiber axis produces two sidebands with the same polarization state but orthogonal, with respect to the polarization of the pump beam. PMI generation was extensively studied and experimentally demonstrated in standard optical fibers in the past [11–13]. In PCFs, PMI was experimentally demonstrated for the first time by Kruhlak et al. [14] in a PCF with normal dispersion at the pump wavelength and pumping near the zero dispersion wavelength (ZDW). In recent works, we have reported experimental PMI generation in all-normal dispersion (ANDi) fibers [15].

Filling the holes with optical materials has been demonstrated to be a useful postfabrication technique for changing the dispersion properties of PCFs. Additionally, it enables fine tuning of the chromatic dispersion, by using the sensitivity of the filling material with external parameters, as, for example, temperature. Several applications related to the nonlinear properties of liquid-filled PCFs have been reported [16–18]. In particular, we demonstrated wide frequency tuning of strong PMI sidebands in ethanol-filled PCFs [19].

Here, we report theoretical simulations and a full experimental study of PMI effect, investigated in a collection of air-filled, ethanol-filled, and heavy water (D_2O)-filled PCFs with different structural parameters and chromatic dispersion characteristics. We used the energy conservation and phase matching condition relations to explore the dispersion and birefringence contributions to PMI frequency shift in ANDi fibers and in fibers with one or two ZDWs.

2. Theoretical Modelling

In the context of optical fibers, PMI is a nonlinear parametric process, in which two photons with linear polarization oriented to one of the principal fiber axes are annihilated to give rise to two new photons with different frequencies, same polarization state, but orthogonal, with respect to pump photons. PMI is generated in singlemode optical fibers through coherent coupling of the two polarized eigenmodes HE_{11x} and HE_{11y}, whose propagation factors are slightly different due to unintentional residual linear birefringence present in the fiber. Depending on the polarization of pump photons, two PMI processes can develop. In the first scenario, the polarization of pump photons is aligned with the slow axis of the fiber, and the PMI photons are generated with polarization oriented to the fast axis. This process is called PMI–SF. In the opposite case, PMI–FS, pump photons are polarized along the fast axes, and the generated photons are polarized along the slow axes.

PMI–SF can produce two sidebands widely spaced and detuned from the pump, even with low pump power. In contrast, PMI–FS presents a power threshold $P_{th} = 3 \cdot \Delta n \cdot A_{eff} / 2 \cdot n_2$ that needs to be reached before PMI generation, where Δn is the modal birefringence, A_{eff} is the modal effective area and n_2 is the fiber nonlinear refractive index. PMI–FS features are somehow more complex, due to the pump power requirements. If $P < P_{th}$, PMI is not generated. When $P_{th} < P < 2P_{th}$, PMI gain shows a maximum value at the zero detuning with no sidebands generation. Only with $P > 2P_{th}$ are sidebands generated close to pump, but with gain remaining different from zero at zero detuning.

Spectral position of PMI sidebands can be calculated from the energy conservation and net phase mismatch relation. The phase–matching condition for both PMI process are described by the following phase equation [9],

$$2\beta_P - \beta_{AS} - \beta_S \pm \frac{\Delta n \cdot (\omega_{AS} + \omega_S)}{c} + \frac{2}{3}\gamma P = 0 \qquad (1)$$

where β_P, β_{AS} and β_S are the propagation factors of the fundamental mode at the pump wavelength, anti-Stokes and Stokes wavelengths, respectively; ω_{AS} and ω_S are the frequencies of the generated photons; c is the speed of light in the vacuum; P is the pump power and γ is the nonlinear coefficient, defined as $\gamma = n_2\, k_p / A_{eff}$, where k_p is the pump wavenumber. In Equation (1) the upper sign stands for PMI–SF, while the lower sign stands for PMI–FS. It can be seen in Equation (1) that both chromatic dispersion and birefringence contribute to the linear phase term. Therefore, PMI frequency shift is strongly affected by both fiber parameters.

We study theoretically the generation of PMI–SF in solid-core PCF with holes filled with air, with ethanol and with heavy water. Figure 1 shows a simple scheme of a solid-core

PCF cross-section showing the structural parameters (i.e., hole diameter d and pitch Λ). First, the linear properties of the fibers were calculated using a fully-vector method [20]. Then, Stokes and anti-Stokes wavelengths were obtained from Equation (1) and the energy conservation condition. Simulations of the PMI–FS process are not included, because it was not investigated in the experiments due to the existence of a power threshold.

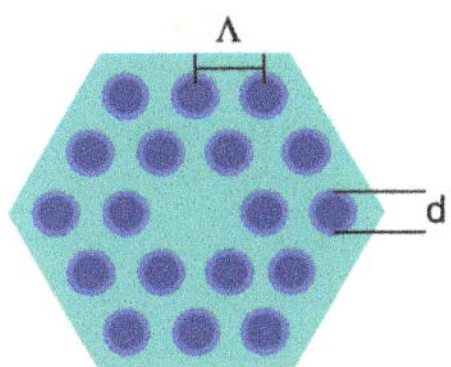

Figure 1. Scheme of cross-section of solid-core photonic crystal fibers (PCF).

The chromatic dispersion of air-filled PCF can be tailored just by changing the fiber structural parameters. Several combinations of d and Λ can be suitable/nm for attaining given dispersion requirements. Figure 2a shows the theoretical calculation of chromatic dispersion for air-filled PCF as a function of the optical wavelength and the fiber pitch. A fixed value of $d/\Lambda = 0.36$ was used in these simulations. The blue level curve in Figure 2a indicates zero dispersion. For this specific air-filling fraction, we can distinguish two different dispersion regimes: (i) when $\Lambda < 1.8$ μm the dispersion presents an ANDi profile, and (ii) when $\Lambda > 1.8$ μm the dispersion profile shows at least one ZDW. In the first regime, fibers with Λ approaching 1.8 μm show low dispersion values in the wavelength range under study, increasing significantly as Λ becomes smaller.

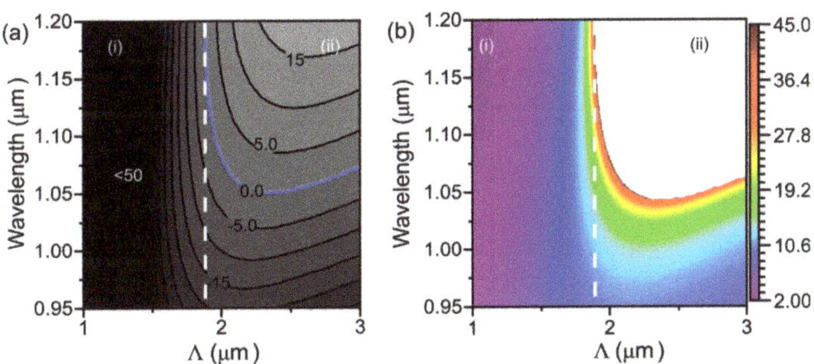

Figure 2. (**a**) Chromatic dispersion as a function of optical wavelength and fiber pitch for air-filled PCFs. Numbers indicates dispersion levels in ps/km·nm. White dashed line indicates pitch of 1.8 μm. (**b**) Frequency shift map of polarization modulation instability (PMI–SF); color bars are detuning frequencies values in THz. Parameters used in the simulations: $d/\Lambda = 0.36$, $P = 3$ kW.

Figure 2b shows a frequency shift map of the PMI–SF process as a function of the optical pump wavelength and fiber pitch. Additional parameters used in PMI–SF simulation are the following: pump power of 3 kW, phase birefringence $\Delta n = 1 \times 10^{-6}$ and nonlinear refractive index $n_2 = 2.7 \times 10^{-20}$ m^2/W. In the case of ANDi fibers, i.e., dispersion region (i), PMI–SF is generated in all cases. For low Λ values, the frequency shift is relatively small, and it shows little dependence on the pump wavelength. The frequency shift increases as the fiber pitch increases, and/or the dispersion at the pump wavelength decreases. In dispersion region (ii), the result is more complex. For Λ slightly above 1.8 μm and pump wavelengths in which the dispersion is small and near to zero (pump wavelength approaching ZDW), PMI–SF experiences a large frequency shift. Additionally, there is a forbidden region where no phase matching solutions for PMI–SF can exist. PMI generated in fibers with one ZDW present a spectral turning point, beyond which the PMI phase matching condition is no longer fulfilled. This occurs when the dispersion mismatch cannot compensate for the contribution of nonlinear mismatch and residual birefringence of the fiber.

The dispersion properties change after filling the holes of PCFs with material different from air, due to the increase of refractive index in the fiber microstructure and the dispersion properties of the new material. For our experiments, we chose ethanol and heavy water, due their suitable refractive index and low absorption in the infrared region. Refractive index characteristics of both liquids used for the theoretical simulations are taken from [21]. Their wavelength dispersion was taken into account in the calculations Figures 3a and 4a show dispersions maps calculated for fibers filled with ethanol and D_2O, respectively. A value of $d/\Lambda = 0.56$ was chosen for D_2O-filled fibers, and $d/\Lambda = 0.58$ for ethanol-filled fibers. These parameters correspond to the mean values of structural parameters of fibers used in the experiments. In both cases, the zero dispersion level occurs for larger wavelengths and larger pitch lengths, compared to air-filled fibers. Figures 3b and 4b show the corresponding frequency shift maps for PMI–SF produced in ethanol-filled and D_2O-filled fibers, respectively. As in the previous case, larger frequency shift occurs in fibers that exhibit low dispersion at the chosen pump wavelength. The shift of zero dispersion that ethanol-filled and D_2O-filled fibers exhibit also causes the movement of the forbidden region of PMI to longer wavelengths and larger pitch lengths.

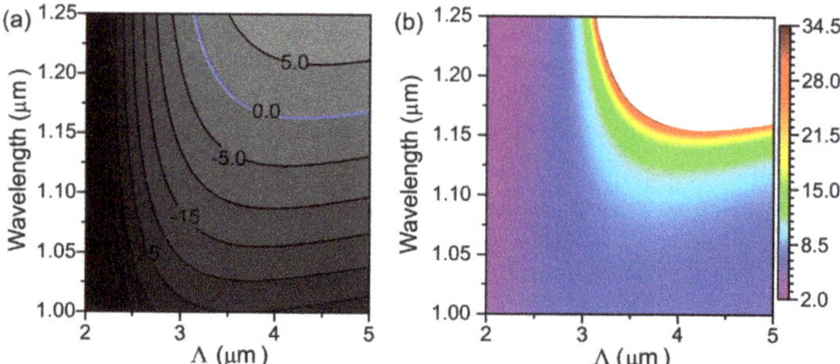

Figure 3. (**a**) Chromatic dispersion as a function of optical wavelength and fiber pitch for ethanol-filled PCFs. Numbers indicates dispersion levels in ps/km·nm. (**b**) Frequency shift map of PMI–SF; color bars are detuning frequencies values in THz. Parameters used in the simulations: $d/\Lambda = 0.58$, $P = 3$ kW.

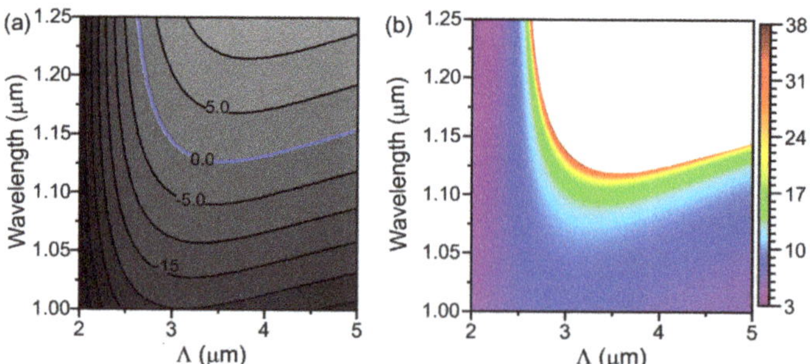

Figure 4. (**a**) Chromatic dispersion as a function of optical wavelength and fiber pitch for D_2O-filled PCFs. Numbers indicates dispersion levels in ps/km·nm. (**b**) Frequency shift map of PMI–SF; color bars are detuning frequencies values in THz. Parameters used in the simulations: $d/\Lambda = 0.56$, $P = 3$ kW.

3. Characteristics of the PCFs

We carried out several experiments to study PMI generation with sub-nanosecond pump pulses in conventional PCFs with air holes and in PCFs filled with different optical liquids. A collection of fibers was made of silica, following the stack-and-draw technique. The PCFs comprise a microstructure with triangular lattice of air holes with a missing hole at the center of the microstructure that acts as the fiber core. After the fibers were fabricated, some of them were infiltrated with ethanol or with D_2O through capillarity force and gas pressure. All fibers were singlemode at the experimental wavelength range. Physical and waveguiding characteristics of the fibers are described in the following sections.

3.1. Air-Filled PCFs

Air-filled PCFs were designed and fabricated to present a convex-ANDi profile in a broad wavelength range. Figure 5a shows scanning electronic microscope (SEM) images of the cross section of the fibers used in the experiments. The structural parameters of the fibers were obtained from the SEM images and are summarized in Table 1. Chromatic dispersion characteristics and effective area were calculated by the method described in [20]. Figure 5b shows the chromatic dispersion as function of wavelength of the four PCFs. Table 1 also includes dispersion values and effective area of the fundamental mode at 1064 nm (i.e., the experimental pump wavelength). In all fibers, the maximum of the curve, i.e., the minimum dispersion wavelength (MDW), is close to the pump wavelength and has a relatively small dispersion value, at 1064 nm.

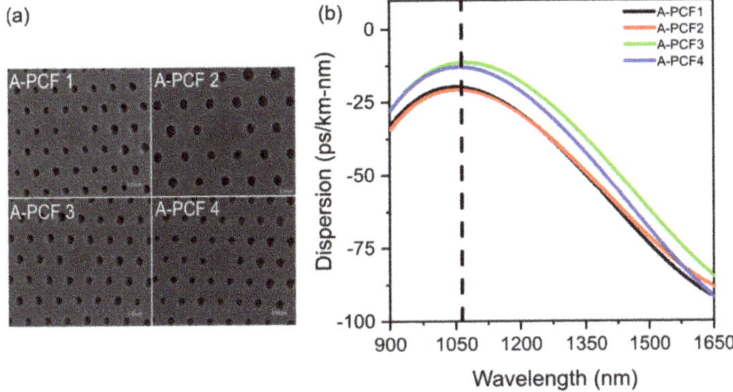

Figure 5. (**a**) SEM images of air-filled PCFs cross-section. (**b**) Chromatic dispersion of air-filled PCFs. Dashed vertical black line indicates the experimental pump wavelength.

Table 1. Structural parameters, dispersion and effective mode area at 1064 nm of air-filled PCFs.

Fiber	Λ (μm)	*d*/Λ	Dispersion (ps/km·nm)	A_{eff} (μm^2)
A-PCF1	1.57	0.35	−19.9	7.2
A-PCF2	1.59	0.35	−20.8	7.5
A-PCF3	1.60	0.36	−11.5	6.7
A-PCF4	1.57	0.37	−13.1	6.5

3.2. Ethanol-Filled PCF

A collection of PCFs with the appropriate dispersion characteristics for PMI generation with the microstructure holes filled with ethanol were designed and fabricated. Figure 6a shows their SEM images taken before infiltration, and Table 2 gives the structural parameters of the fibers. The guiding characteristics were calculated, taking into account the refractive index of ethanol, including its wavelength dispersion [21]. Figure 6b shows the dispersion profile of the six fibers investigated. All of them exhibit normal dispersion at 1064 nm. According with their dispersion characteristics, different types of fiber were obtained. Three fibers showed an ANDi profile (E-PCF1, E-PCF2 and E-PCF3), with different values of dispersion at the MDW and increasing values of dispersion at 1064 nm. Two fibers (E-PCF4, E-PCF5) showed a dispersion profile that shows two ZDWs, the first near to 1200 nm and second at 1600 nm, quite far from the pump wavelength. These two fibers have very similar characteristics, although not identical. The results of both PCFs have been included to show that PMI frequency shift can be very sensitive to the fiber properties. Finally, we included in this study one PCF (E-PCF6) with a dispersion profile showing just one ZDW (≈1117 nm) in the vicinity of the pump wavelength and a small value of normal dispersion at the pump wavelength. It is noteworthy that the chromatic dispersion of all these six fibers having air in the holes are anomalous at the pump wavelength, with the ZDW quite below the pump wavelength.

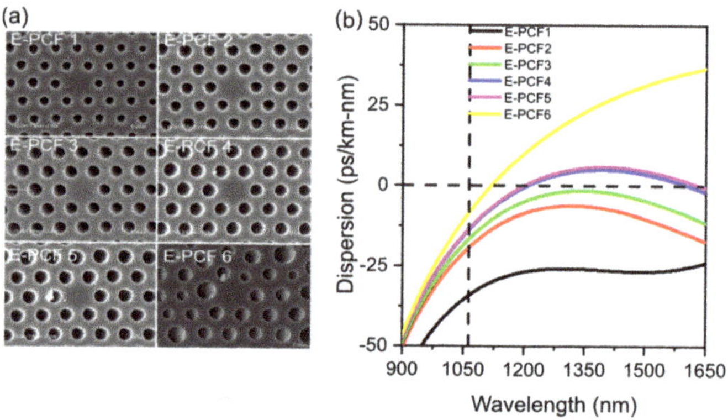

Figure 6. (**a**) SEM images of ethanol-filled PCFs cross-section. (**b**) Chromatic dispersion of PCFs with holes filled ethanol. Dashed vertical black line indicates the experimental pump wavelength.

Table 2. Structural parameters, dispersion and effective mode area at 1064 nm of ethanol-filled PCFs.

Fiber	Λ (μm)	d/Λ	Dispersion (ps/km·nm)	ZDW (nm)	A_{eff} (μm²)
E-PCF1	2.66	0.43	−34.1	-	20.1
E-PCF2	2.76	0.54	−19.1	-	12.3
E-PCF3	2.83	0.57	−16.1	-	11.6
E-PCF4	2.90	0.59	−13.7	1208	11.2
E-PCF5	2.92	0.60	−13.4	1200	11.2
E-PCF6	3.97	0.75	−8.4	1117	13.1

3.3. D_2O-Filled PCF

Figure 7a shows SEM images of the fibers that were infiltrated with D_2O. Table 3 gives their structural parameters as well as the chromatic dispersion and effective area at 1064 nm. The guiding properties were calculated, taking into account the refractive index of D_2O and its wavelength dispersion [21]. Figure 7b shows their dispersion characteristics. All of them exhibited normal dispersion at pump wavelength. One PCF (D-PCF1) showed an ANDi profile, and the dispersion profile of the rest of PCFs showed one ZDW at different values of wavelength, from 1092 nm to 1140 nm, and increasing dispersion values at 1064 nm, from −9.1 ps/nm km to −3.9 ps/nm km.

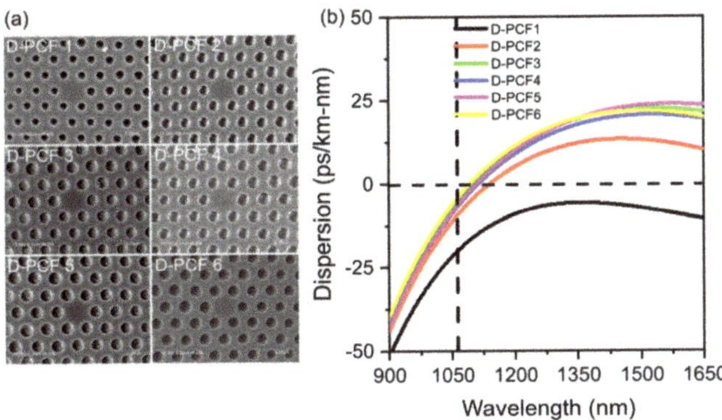

Figure 7. (**a**) SEM images of D_2O-filled PCFs cross-section. (**b**) Chromatic dispersion of PCFs with holes filled with D_2O. Dashed vertical black line indicates the experimental pump wavelength.

Table 3. Structural parameters, dispersion and effective mode area at 1064 nm of D_2O-filled PCFs.

Fiber	Λ (μm)	d/Λ	Dispersion (ps/km·nm)	ZDW (nm)	A_{eff} (μm^2)
D-PCF1	2.68	0.44	−19.9	-	15.0
D-PCF2	2.78	0.54	−9.1	1140	10.9
D-PCF3	2.86	0.59	−5.3	1102	9.9
D-PCF4	2.92	0.56	−6.6	1112	11.2
D-PCF5	2.95	0.58	−6.2	1107	10.8
D-PCF6	2.76	0.62	−3.9	1091	9.0

4. Experimental Setup

Figure 8 shows the experimental arrangement scheme. The pump laser is a passively Q-switched Nd: YAG microchip laser (TEEM Photonics SNP 20F-100) that emits pulses at 1064.5 nm of 700 ps duration (FWHM), few kW of peak power, and a repetition rate of 19.1 kHz. The laser emission is linearly polarized with a polarization extinction ratio (PER) of 32 dB. The laser beam was delivered into the fiber under test (FUT). The length of the fibers used in the experiments was about ≈1 m. A half-wave plate (HWP) was used to rotate the polarization plane of the pump beam. The light exiting the optical fiber was collected and analyzed with an optical spectrum analyzer (OSA).

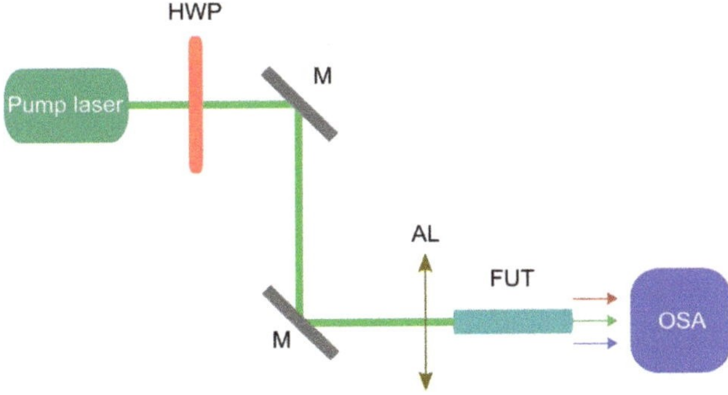

Figure 8. Experimental arrangement. Half-wave plate (HWP). Mirror (M). Aspherical lens (AL). Fiber under test (FUT). Optical spectrum analyzer (OSA).

5. Experimental Results and Discussion

Light spectra from the fibers' output were recorded when the polarization orientation of the pump was aligned to the main fiber axes. Figure 9 shows the spectrum obtained from fiber A-PCF1 for two orthogonal polarization orientations and the same pump power. The black line corresponds to pump polarization, matching the slow axis of the fiber. Two strong sidebands, centered at 991 nm and 1148 nm, are generated in the fiber. These two sidebands refer to anti-Stokes and Stokes photons generated via the PMI–SF process. The blue line shows the spectrum obtained after the pump polarization orientation was rotated by 90° (i.e., fast-axis pumping). The strong sidebands described above are not present anymore, and the rest of the spectrum remains practically the same. PMI–FS was not observed in the experiment, as the pump power level (≈2.4 kW) injected into the fiber was below the power threshold required for PMI–FS to occur in this fiber (P_{th} = 6.4 kW). The two secondary sidebands of lower amplitude, centered at 1016 nm and 1116 nm (frequency shift ≈13.2 THz) and shown in both spectra, are produced by Raman scattering (RS).

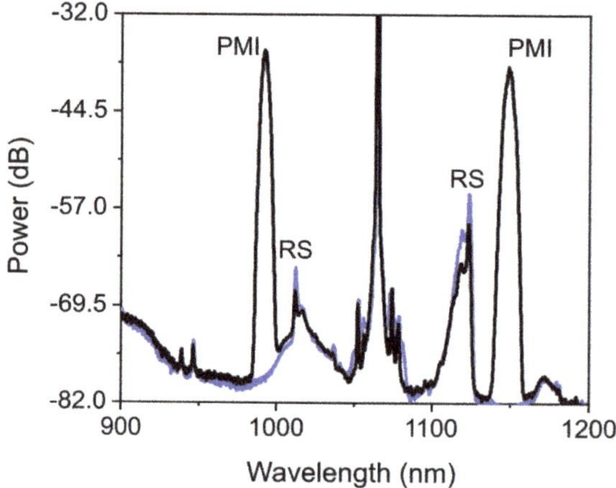

Figure 9. Light spectrum for two orthogonal polarization orientation of the pump. Pump polarization oriented to the fast axis (blue line) and to the slow axis (black line). Both spectra were recorded at pump peak power of 2.4 kW.

The polarization of the two strong bands shown in Figure 9 was experimentally analyzed with a bulk linear polarizer (LP). First, the HWP was adjusted to optimize the amplitude of the bands. The HWP was kept fixed during the experiment at this orientation. Then, the LP was inserted after the fiber end, and it was rotated to obtain best transmission of the bands. At this LP orientation, the amplitude of the PMI–SF bands was the maximum, while the amplitude of the rest of the spectral components (i.e., the residual pump and the RS bands) were attenuated a few tens of dB. When the LP was rotated 90° degrees from the earlier orientation, the amplitude of both PMI bands decreased into the noise level, while the rest of spectral components experienced an amplitude increase of a few tens of dB. This behavior confirms that the polarization state of PMI–SF bands is orthogonal to RS and the remaining pump light.

The generation of PMI–SF in the PCFs described in the Section 3 was investigated. Figure 10 shows the resulting optical spectra from each sample. Notice that high orders of PMI–SF were generated in some fibers. Table 4 summarizes the spectral position (and frequency shift) of first-order PMI–SF bands generated in the fibers.

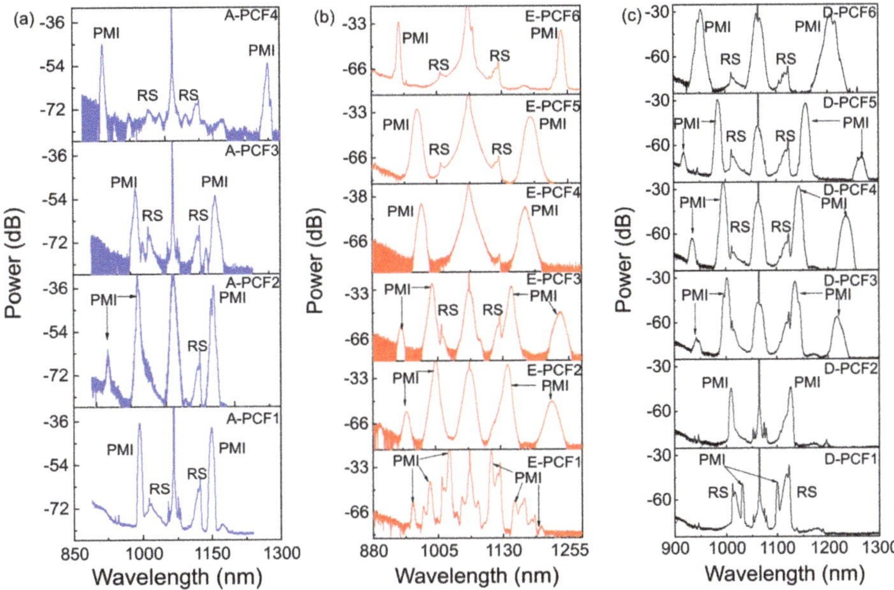

Figure 10. PMI spectra generated in (**a**) air-filled PCFs (blue line), (**b**) ethanol-filled PCFs (red line) and (**c**) D_2O-filled PCFs (black line).

Table 4. Wavelength and frequency shift of PMI–SF bands generated in air-filled, ethanol-filled and D_2O-filled PCFs. Modal birefringence.

Filling Substance	Fiber	Anti-Stokes (nm)	Stokes (nm)	Frequency Shift (THz)	Birefringence
Air	A-PCF1	991.4	1148.3	20.57	1.64×10^{-5}
	A-PCF2	988.0	1153.8	21.81	1.72×10^{-5}
	A-PCF3	985.0	1158.2	22.80	9.56×10^{-6}
	A-PCF4	914.6	1272.9	46.14	6.41×10^{-5}
Ethanol	E-PCF1	1025.4	1106.0	10.57	6.28×10^{-6}
	E-PCF2	1000.0	1138.2	18.25	1.03×10^{-5}
	E-PCF3	993.7	1146.6	20.18	1.05×10^{-5}
	E-PCF4	974.1	1173.7	26.22	1.60×10^{-5}
	E-PCF5	965.2	1185.8	28.83	1.80×10^{-5}
	E-PCF6	931.7	1242.0	40.28	2.24×10^{-5}
D_2O	D-PCF1	1031.0	1100.0	9.10	1.55×10^{-6}
	D-PCF2	1009.2	1126.7	15.56	1.91×10^{-6}
	D-PCF3	1002.0	1136.0	17.74	7.0×10^{-7}
	D-PCF4	995.0	1144.0	19.58	3.40×10^{-6}
	D-PCF5	985.6	1157.8	22.71	3.79×10^{-6}
	D_PCF6	952.8	1206.1	33.09	5.21×10^{-6}

As it will be shown later in detail, in the case of ANDi PCFs with convex dispersion profile, the largest PMI shift was attained when the pump wavelength was close to the MDW, and the dispersion at the pump wavelength was as small as possible. This is shown experimentally in Figure 10, where the largest frequency shift (≈46 THz) occurs in fiber A-PCF4, with anti-Stokes and Stokes bands centered at 914 nm and 1273 nm, respectively. When the pump wavelength is far from the MDW, so that the fiber dispersion at the pump wavelength is larger, the frequency shift becomes smaller. As a representative example, fiber D-PCF1 generated PMI–SF bands centered at 1031 and 1100 nm, which corresponds to a frequency shift of 9 THz, which was the smallest obtained in the experiments.

The frequency shift of PMI–SF generated in fibers with more conventional dispersion profiles can also be very large when the pump wavelength is close to a ZDW and the dispersion is small. From our experiments, we can investigate the generation of PMI in fibers with one ZDW with different chromatic dispersion at the pump wavelength. For example, the results obtained from the sequence of fibers E-PCF4, E-PCF5 and E-PCF6, with dispersion values decreasing from -13.7 ps/(nm·km) to -8.4 ps/(nm·km), show PMI frequency shift clearly increasing as the dispersion value decreases, from $\approx$26 THz in the case of E-PCF4 to $\approx$40 THz for E-PCF6.

A similar trend can be seen if we consider the series of fibers filled with D_2O, with the exception of D-PCF1 that exhibits an ANDi profile. The largest frequency shift ($\approx$33 THz) occurred in fiber D-PCF6, which is the fiber with lowest dispersion, while the fiber D-PCF2 with largest dispersion showed the shortest frequency shift ($\approx$15 THz)

It is worth noting that, in some fibers with similar values of dispersion at 1064 nm, PMI–SF produces bands with quite different frequency shifts. For example, the frequency shift in ANDi fibers A-PCF1 and D-PCF1 is $\approx$20 THz and $\approx$9 THz, respectively. Similarly, it happens for some fibers filled with D_2O. Residual phase birefringence is the origin of this apparent discrepancy. As stated in Equation (1), phase birefringence also plays a major role in the phase matching condition and adds positively to the frequency shift. The larger the birefringence, the more it contributes to the frequency shift.

The wavelengths at which the bands were generated varied accordingly with the dispersion characteristics and residual birefringence of each fiber. In general terms, experimental results can be described appropriately with theoretical calculations of PMI–SF wavelengths, taking into account the fibers' properties. Fiber birefringence values stated in Table 4 were obtained from fitting the theoretical calculations of frequency shift to the experimental results. The resulting values of birefringence range from 10^{-5} to 10^{-7}, which are compatible with residual phase birefringence values in solid core PCFs. Figure 11 shows the theoretical results of PMI–SF frequency shift as function of pump wavelength for the different fibers, along with the experimental data.

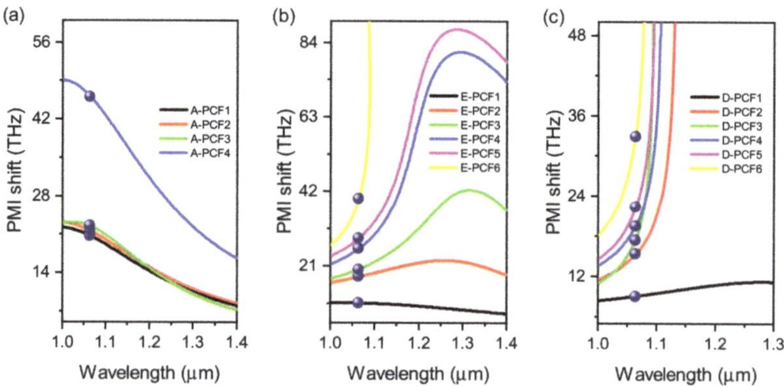

Figure 11. PMI–SF frequency shift as function of pump wavelength calculated for (**a**) air-filled PCFs, (**b**) ethanol-filled PCFs and (**c**) D_2O-filled PCFs. Dots are experimental data. Peak power of 2.5 kW for air-filled PCFs and 3 kW for both ethanol-filled PCFs and D_2O-filled PCFs.

As mentioned before, it is shown that the frequency shift in fibers with one ZDW shows a phase matching turning point at a wavelength close to the ZDW, so that PMI–SF can no longer occur for pump wavelengths above it. Large frequency shifts can be attained by pumping near the wavelength at which the turning point takes place, as it happens for fibers E-PCF6 and D-PCF6. Operating near the turning point might have also some

drawbacks: the large frequency shift slope is likely to cause the generation Additionally, the parametric wavelengths are very sensitive to small fluctu geometry, which, in practice, can lead to unwanted broadening of the ge

In the case of PCFs with two relatively close ZDWs and ANDi PCF convex dispersion profile, the largest frequency shift is attained when the p is close to the MDW. Phase matching can occur regardless of the pump w first type, even under anomalous dispersion pumping. However, it is imp that additional four-wave mixing processes, in particular, scalar FWM instability, can happen in fibers with dispersion profiles showing one or t they are forbidden in ANDi fibers.

Finally, the strong dependence of PMI frequency shift on phase birefri in the theoretical simulations. For instance, comparing fibers A-PCF3 ar rather similar dispersion characteristics, they show quite different PMI (see Figure 11a), due to the different contributions of birefringence to the

6. Conclusions

In summary, we have reported a detailed study regarding PMI gener tion of PCFs. Air-filled, ethanol-filled and D_2O-filled PCFs featuring diff dispersion characteristics have been investigated using long pump pulses a been shown that PMI frequency shift can be very large in ANDi fibers with values when they are pumped close to the MDW. The largest frequency s our experiments for ANDi fibers was 46 THZ in a PCF with -13.1 ps/nm·k at the pump wavelength and phase birefringence of 6.4×10^{-5}. Large fr also observed in PCFs with dispersion profiles with one ZDW (or two Z pumped near the ZDW. Frequency shifts of 40 THz and 32.9 THz were ethanol-filled PCF (E-PCF6) and a D_2O-filled PCF (D-PCF6), respectively fibers were pumped near the PMI phase matching turning point. Phase mat and energy conservation were used to theoretically investigate the freque PMI–SF process in the fibers investigated experimentally, showing good a the experimental data.

In the present work, PMI has been investigated in a collection of fib wide range of structural parameters and dispersion regimes. We hope reported here will be useful for future works that might involve optical f PMI generation.

Author Contributions: Conceptualization, A.D., A.L.-T. and M.V.A.; methodology, software, E.S. and A.L.-T.; formal analysis, A.L.-T., A.D. and M.V.A.; investigat A.D.; writing, A.L.-T., A.D. and M.V.A.; supervision, M.V.A.; project administratio ing acquisition, M.V.A. and A.D. All authors have read and agreed to the publi the manuscript.

Funding: Ministerio de Ciencia e Innovación and Fondo Europeo de Desarrollo Re 104276RB-I00); Generalitat Valenciana (PROMETEO/2019/048 and IDIFEDER/2020 pean Commission (H2020-MSCA-RISE-2019-872049).

Conflicts of Interest: The authors declare no conflict of interest.

References

1. Dudley, J.M.; Genty, G.; Coen, S. Supercontinuum generation in photonic crystal fiber. *Rev. Mod. P* 1135–1184. [CrossRef]
2. Huang, C.; Liao, M.; Bi, W.; Li, X.; Hu, L.; Zhang, L.; Wang, L.; Qin, G.; Xue, T.; Chen, D.; et al. Ultraflat, broadba coherent supercontinuum generation in all-solid microstructured optical fibers with all-normal dispersion. *Photon* 601–608. [CrossRef]
3. Knight, J.C. Photonic crystal fibres. *Nature* **2003**, *424*, 847–851. [CrossRef] [PubMed]
4. Zhang, L. *Ultra-Broadly Tunable Light Sources Based on the Nonlinear Effects in Photonic Crystal Fibers*; Springer: Berl Germany, 2016; ISBN 978-3-662-48359-6.

Cruz, F.C. Optical frequency combs generated by four-wave mixing in optical fibers for astrophysical spectrometer calibration and metrology. *Opt. Express* **2008**, *16*, 13267–13275. [CrossRef] [PubMed]

Gottschall, T.; Meyer, T.; Jauregui, C.; Schmitt, M.; Popp, J.; Limpert, J.; Tünnermann, A. Four-wave mixing based light sources for real-world biomedical applications of coherent Raman microscopy. *Proc. SPIE* **2016**, *9712*, 971202.

Gottschall, T.; Meyer, T.; Baumgartl, M.; Jauregui, C.; Schmitt, M.; Popp, J.; Limpert, J.; Tünnermann, A. Fiber-based light sources for biomedical applications of coherent anti-Stokes Raman scattering microscopy. *Laser Photonics Rev.* **2015**, *9*, 435–451. [CrossRef]

Velázquez-Ibarra, L.; Díez, A.; Silvestre, E.; Andrés, M.V.; Lucio, J.L. Modeling spectral correlations of photon-pairs generated in liquid-filled photonic crystal fibers. *J. Opt.* **2020**, *22*, 075203. [CrossRef]

Agrawal, G.P. *Nonlinear Fiber Optics*, 5th ed.; Academic Press: Boston, MA, USA, 2013; ISBN 15575837.

Kudlinski, A.; Bendahmane, A.; Labat, D.; Virally, S.; Murray, R.T.; Kelleher, E.J.R.; Mussot, A. Simultaneous scalar and cross-phase modulation instabilities in highly birefringent photonic crystal fiber. *Opt. Express* **2013**, *21*, 8437–8443. [CrossRef] [PubMed]

Murdoch, S.G.; Leonhardt, R.; Harvey, J.D. Polarization modulation instability in weakly birefringent fibers. *Opt. Lett.* **1995**, *20*, 866–868. [CrossRef] [PubMed]

Millot, G.; Seve, E.; Wabnitz, S.; Haelterman, M. Observation of induced modulational polarization instabilities and pulse-train generation in the normal-dispersion regime of a birefringent optical fiber. *J. Opt. Soc. Am. B* **1998**, *15*, 1266–1277. [CrossRef]

Kockaert, P.; Haelterman, M.; Pitois, S.; Millot, G. Isotropic polarization modulational instability and domain walls in spun fibers. *Appl. Phys. Lett.* **1999**, *75*, 2873–2875. [CrossRef]

Kruhlak, R.J.; Wong, G.K.; Chen, J.S.; Murdoch, S.G.; Leonhardt, R.; Harvey, J.D.; Joly, N.Y.; Knight, J.C. Polarization modulation instability in photonic crystal fibers. *Opt. Lett.* **2006**, *31*, 1379–1381. [CrossRef] [PubMed]

Loredo-Trejo, A.; López-Diéguez, Y.; Velázquez-Ibarra, L.; Díez, A.; Silvestre, E.; Estudillo-Ayala, J.M.; Andrés, M.V. Polarization modulation instability in all-normal dispersion microstructured optical fibers with quasi-continuous pump. *IEEE Photonics J.* **2019**, *11*, 1–8. [CrossRef]

Ott, J.R.; Heuck, M.; Agger, C.; Rasmussen, P.D.; Bang, O. Label-free and selective nonlinear fiber-optical biosensing. *Opt. Express* **2008**, *16*, 20834–20847. [CrossRef] [PubMed]

Frosz, M.H.; Stefani, A.; Bang, O. Highly sensitive and simple method for refractive index sensing of liquids in microstructured optical fibers using four-wave mixing. *Opt. Express* **2011**, *19*, 10471–10484. [CrossRef] [PubMed]

Velázquez-Ibarra, L.; Díez, A.; Silvestre, E.; Andrés, M.V. Tunable four-wave mixing light source based on photonic crystal fibers with variable chromatic dispersion. *J. Lightwave Technol.* **2019**, *37*, 5722–5726. [CrossRef]

Loredo-Trejo, A.; Díez, A.; Silvestre, E.; Andrés, M.V. Broadband tuning of polarization modulation instability in microstructured optical fibers. *Opt. Lett.* **2020**, *45*, 4891–4894. [CrossRef] [PubMed]

Silvestre, E.; Pinheiro-Ortega, T.; Andrés, P.; Miret, J.J.; Ortigosa-Blanch, A. Analytical evaluation of chromatic dispersion in photonic crystal fibers. *Opt. Lett.* **2005**, *30*, 453–455. [CrossRef] [PubMed]

Kedenburg, S.; Vieweg, M.; Gissibl, T.; Giessen, H. Linear refractive index and absorption measurements of nonlinear optical liquids in the visible and near-infrared spectral region. *Opt. Mater. Express* **2012**, *2*, 1588–1611. [CrossRef]

MDPI
St. Alban-Anlage 66
4052 Basel
Switzerland
Tel. +41 61 683 77 34
Fax +41 61 302 89 18
www.mdpi.com

Crystals Editorial Office
E-mail: crystals@mdpi.com
www.mdpi.com/journal/crystals

www.ingramcontent.com/pod-product-compliance
Lightning Source LLC
LaVergne TN
LVHW070614170726
843515LV00004B/77

* 9 7 8 3 0 3 6 5 1 6 1 4 1 *